# Detection of Drugs and Their Metabolites in Oral Fluid

# Emerging Issues in Analytical Chemistry

Series Editor
## Brian F. Thomas

Co-published by Elsevier and RTI Press, the Emerging Issues in Analytical Chemistry series highlights contemporary challenges in health, environmental, and forensic sciences being addressed by novel analytical chemistry approaches, methods, or instrumentation. Each volume is available as an e-book, on Elsevier's ScienceDirect, and via print. Series editor Dr. Brian F. Thomas continuously identifies volume authors and topics; areas of current interest include identification of tobacco product content prompted by regulations of the Family Tobacco Control Act, constituents and use characteristics of e-cigarettes and vaporizers, analysis of the synthetic cannabinoids and cathinones proliferating on the illicit market, medication compliance and prescription pain killer use and diversion, and environmental exposure to chemicals such as phthalates, endocrine disrupters, and flame retardants. Novel analytical methods and approaches are also highlighted, such as ultraperformance convergence chromatography, ion mobility, in silico chemoinformatics, and metallomics. By highlighting analytical innovations and new information, this series furthers our understanding of chemicals, exposures, and societal consequences.

# Detection of Drugs and Their Metabolites in Oral Fluid

Robert M. White, Sr.
Retired from RTI International, Research Triangle Park, NC;
RMW Consulting, Inc., Naples, FL, USA

Christine M. Moore
Immunalysis Corporation (now a part of Abbott), Pomona, CA, USA

Elsevier
Radarweg 29, PO Box 211, 1000 AE Amsterdam, Netherlands
The Boulevard, Langford Lane, Kidlington, Oxford OX5 1GB, United Kingdom
50 Hampshire Street, 5th Floor, Cambridge, MA 02139, United States

Published in cooperation with RTI Press at RTI International, an independent, nonprofit research institute
that provides research, development, and technical services to government and commercial clients worldwide
(www.rti.org). RTI Press is RTI's open-access, peer-reviewed publishing channel. RTI International is a trade
name of Research Triangle Institute.

**British Library Cataloguing-in-Publication Data**
A catalogue record for this book is available from the British Library

**Library of Congress Cataloging-in-Publication Data**
A catalog record for this book is available from the Library of Congress

ISBN: 978-0-12-814595-1

For Information on all Elsevier publications
visit our website at https://www.elsevier.com/books-and-journals

  **Working together
to grow libraries in
developing countries**

www.elsevier.com • www.bookaid.org

*Publisher:* John Fedor
*Acquisition Editor:* Kathryn Morissey
*Editorial Project Manager:* Amy Clark
*Production Project Manager:* Paul Prasad Chandramohan
*Designer:* Mathew Limbert

Typeset by MPS Limited, Chennai, India

**Christine M. Moore**

I should like to thank my sister, Anne, for her constant support and my colleagues at Immunalysis for all their exemplary research and scientific approach to all aspects of analyzing drugs in oral fluid.

**Robert M. White, Sr.**

To all my family for tolerating endless stacks of paper throughout the writing and publication of six books.

# CONTENTS

# PREFACE

According to Irwin Mandel, saliva "lacks the drama of blood, the sincerity of sweat and the emotional appeal of tears."[*] However, oral fluid, of which saliva is the main component, has found numerous uses in clinical and forensic toxicology and medicine. Like its counterpart plasma, it contains proteins and polypeptides of diagnostic significance. It is used in the diagnosis of non-microbial disease states such as Sjögren's syndrome, Cushing's disease, and celiac disease. It is useful for diseases that have a specific microorganism as their root cause, such as herpes virus associated with Kaposi sarcoma; *Helicobacter pylori* associated with gastritis, peptic ulcer, and possible stomach cancer; and human immunodeficiency virus (HIV). High quality DNA (deoxyribonucleic acid) may be extracted from oral fluid in usable quantity. However, at the time of this writing, the main value of oral fluid lies in the analysis of small organic molecules including, but not limited to, hormones such as cortisol and drugs, both therapeutic and illicit. The main purpose of this short book is to identify where oral fluid drug and drug metabolite testing is useful, what the limitations of the tests are, and how to accomplish testing for a variety of specialties.

---

[*]Mandel ID. The diagnostic uses of saliva. *J Oral Pathol Med.* 1990; 19(3):119–25. PubMed ID 2187975.

# ACKNOWLEDGMENTS

Christine M. Moore acknowledges the support of all the staff at Immunalysis during this project. She is an employee of Immunalysis (now a part of Abbott), which manufactures the Quantisal oral fluid collection device as well as numerous immunoassays intended for the analysis of drugs in oral fluid. In addition, Abbott manufactures several point-of-collection rapid tests for oral fluid, including the DDS2 device, also discussed in this book.

Robert M. White, Sr., retired from RTI International August 2017, acknowledges the support of RTI International and the Substance Abuse and Mental Health Services Administration for oral fluid projects and programs referenced in this book.

# CHAPTER 1

# Introduction

## GENERAL

Before considering oral fluid as a matrix for drug and drug metabolite testing, it is worthwhile describing the sources and formation of human oral fluid in order to understand its strong points and limitations.

Saliva, the major component of oral fluid, is produced primarily by three bilateral pairs of salivary glands: parotid, submandibular, and sublingual, as summarized in Table 1.1.[1]

Fig. 1.1 shows the approximate location of each pair of glands.[2]

Minor accessory salivary glands, which are about 450−750 in number, are located on the tongue, the buccal mucosa of the lips, and the palate. They do not have a common duct, and they produce a viscous secrete.[1] Von Ebner's glands (not shown in Fig. 1.1), which are oral exocrine glands that in part secrete lipase, are serous glands that reside at the base of the crypts that surround the circumvallate and foliate papillae on the tongue, just anterior to the posterior third of the tongue.[3]

A healthy individual produces about 500−1500 mL of saliva per day, which is about 0−6 mL per minute, depending highly on whether the individual is awake or asleep and whether the salivary glands are stimulated or not, as illustrated in Table 1.2.[1]

A salivary gland consists of acini and a duct that leads either directly to the oral cavity or to a common duct that leads to the oral cavity. Fig. 1.2 shows that circulating blood exchanges components at the acini and the ductus.[2] In both areas, a capillary bed exists to facilitate the exchange.

In Fig. 1.2, it is most notable that several pathways, not just simple filtration, exist for the transfer of small molecules such as drugs and

Detection of Drugs and Their Metabolites in Oral Fluid. DOI: https://doi.org/10.1016/B978-0-12-814595-1.00001-5

| Table 1.1 Major Salivary Glands With Their Associated Ducts and Secretions | | | |
|---|---|---|---|
| Gland | Duct | Secrete | Major Non-Water Product |
| Parotid | Stenson | Serous | Amylase, proline-rich protein; no mucins |
| Submandibular | van Wharton | Seromucous | Mucins |
| Sublingual | Bartholin | Mucous only | Mucins |

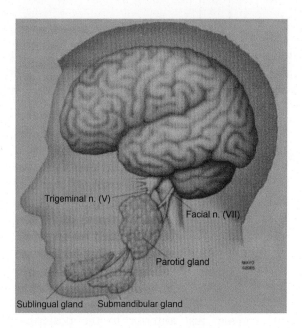

*Figure 1.1 Approximate locations of major human salivary glands.*

| Table 1.2 Percent Contribution of Each Salivary Gland by Stimulation Type | | | | |
|---|---|---|---|---|
| Gland | Sleep | No Stimulation | Mechanical Stimulation[a] | Citric Acid Stimulation[b] |
| Parotid | 0 | 21 | 58 | 45 |
| Submandibular | 72 | 70 | 33 | 45 |
| Sublingual | 14 | 2 | 2 | 2 |
| Minor glands | 14 | 7 | 7 | 8 |

[a]This type of stimulation may be accomplished by chewing on an inert substance such as paraffin.
[b]Also called gustatory stimulation.

drug metabolites from capillary plasma into saliva that is being formed in the salivary gland.

Saliva becomes oral fluid when formed saliva enters the oral cavity and mixes with crevicular fluid, which contains small amounts of normal human plasma constituents such as immunoglobulin G (IgG), oral

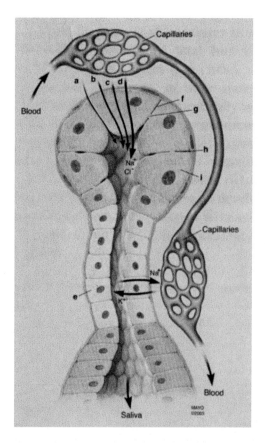

*Figure 1.2 Mechanisms of transport of proteins and ions from plasma into the salivary ducts. a, ultrafiltration; b, active transport or passive diffusion across the cell membrane; c, simple filtration through cell membrane pores; d, transepithelial movement of water along NaCl gradient via channel proteins; e, creation of hypotonic salivary solution via ductal Na⁺ reabsorption; f, acinar cell membrane; g, cell membrane pore; h, intercellular space; i, acinar cell.*

cavity microbes, nasopharyngeal secretions, and any food particles that may be present.

The actions of the acini produce an ultrafiltrate of blood with a composition that is vastly different from plasma, which is the noncellular component of whole blood (as illustrated in Table 1.3) from which the oral fluid was derived.[2,3]

Increased secretion flow rate results in increased pH primarily due to increased bicarbonate. Also, as salivary flow rate increases, saliva sodium ion ($Na^+$) concentration increases due to reduced opportunity for uptake immediately prior to the ductus. A compensatory rise in the

**Table 1.3 Components of Saliva (Normally Measured as Oral Fluid) Compared to Blood Plasma[2]**

| Parameter | Plasma[4] | Whole Human *Resting* Oral Fluid[a] | Whole Human *Stimulated* Oral Fluid[a] |
|---|---|---|---|
| pH (arterial, whole blood except for oral fluid) | 7.35−7.45 (children, adults) | 6.8 (6.2−7.4)[5] | Increased. Please see text. |
|  | 7.31−7.42 (60−90 years) 7.26−7.93(>90 years) | 6.7 (5.6−7.9)[6] |  |
| $Na^+$ (mmol/L; mEq/L) | 137−143 (16−49 years; male) 137−142 (16−49 years; female) | 5 | 20−80 |
|  | 136−143 (50−79 years) |  |  |
| $K^+$ (mmol/L; mEq/L) | 3.8−4.9 (6−79 years) | 22 | 20 |
| Total $Ca^{++}$ (mmol/L) | 9.1−10.4 (20−39 years; male) 9.0−10.1 (20−39 years; female) | 1−4 | 1−4 |
|  | 9.0−10.2 (40−79 years) |  |  |
| $Mg^{++}$ (mmol/L); serum | 0.68−1.07 (>12 years) | 0.2 | 0.2 |
| $Cl^-$ (mmol/L) | 101−106 (12−29 years; male) 100−107 (12−29 years; female) | 15 | 30−100 |
|  | 102−108 (30−79 years) |  |  |
| Total $CO_2$; primarily $HCO_3^-$ (mmol/L) | 19−24 (arterial, 6−79 years) | 5 | 15−80 |
|  | 22−26 (venous, 6−79 years) |  |  |
| $PO_4^{-3}$ (mmol/L) | 2.9−4.7 (16−47 years) | 6 | 4 |
|  | 2.8−4.7 (48−79 years; male) 3.1−4.8 (48−79 years; female) |  |  |
| $SCN^-$ (mmol/L); serum | <0.4 (nonsmokers) | 2.5 | 2 |
|  | <1.2 (smokers) |  |  |
| $NH_3$ (mmol/L) | 15−45 | 6 | 3 |
| $(NH_2)_2CO$ (mmol/L) | 2.9−7.0 (8−19 years; male) 3.3−7.9 (20−39 years; male) 3.5−8.6 (40−59 years; male) 2.7−6.7 (8−59 years; female) | 3.3 | 2−4 |
|  | 3.6−9.2 (60−79 years) |  |  |
| Total protein (g/L) | 68−82 (6−19 years) 65−83 (20−29 years) | 3 | 3 |
|  | 65−78 (30−79 years) |  |  |
| Mucin (g/dL) | Not detected | 0.27 (0/08−0.6) |  |
| [a]From Reference[1] except as noted. | | | |

major counterion, $Cl^-$, occurs due to the increased sodium ions. Increased bicarbonate also means an increased buffering capacity. It has been the authors' experience that even though oral fluid is approximately 99% water, it has an unexpectedly high buffering capacity. It is well worth a laboratory's time to validate the ability of their collection device by ascertaining that it maintains its initial pH in oral fluid from at least 10 separate donors. For such a study, approval from an institutional review board should be obtained. It has been the experience of the authors that vitamin C (ascorbic acid) can be used successfully as a salivary stimulant when large batches of oral fluid (e.g., $\geq 50$ mL) are required. The pH of the mixed oral fluid usually is around 6.6−6.8, not elevated as might be anticipated. If ascorbic acid is used, its influence or lack thereof on analytical methodology needs to be validated.

An oral fluid specimen is one collected from a donor's oral cavity and is a combination of physiological fluids produced primarily by the salivary glands.[7] Whole oral fluid, many times misreferenced as whole saliva or mixed saliva, is a mixture of oral fluids and includes secretions from the minor and major salivary glands and constituents of nonsalivary origin, including gingival crevicular fluid, expectorated nasal and bronchial secretions, blood and blood derivatives from oral wounds, bacteria and bacterial products, viruses, fungi, desquamated epithelial cells and other cellular material, and food particles. Albumin and IgG arise in oral fluid primarily from crevicular fluid. It is notable that crevicular fluid is reduced or absent in edentulous people.[8]

Small molecules such as drugs, drug metabolites, other xenobiotics, and naturally produced hormones commonly are thought to be present in oral fluid due to simple filtration in the sulcus of the oral fluid gland with ion trapping of basic (usually nitrogen-containing drugs and their metabolites) molecules, since the pH of oral fluid usually is lower than that of the arterial blood from which the small molecules are transferred. However, the process by which small molecules in the arterial capillary lumen end up in the lumen of the salivary gland is more complex than often considered. Usually, at least five barriers must be passed for a substance to travel from the vascular lumen to the ductal system where initially formed saliva is transformed into the final product (please see Fig. 1.2 and Table 1.3) that exits the duct.[2]

1. The capillary wall
2. The interstitial space

3. The basal cell membrane of the acinar cell
4. The fluid inside the acinar cell
5. The luminal cell membrane

At least five known factors affect the movement of substances from capillary blood into saliva that will eventually exit the salivary duct into the oral cavity.[2]

1. Molecular mass and size: Mass plays a minor role in regulating diffusion. The diffusion coefficient is inversely proportional to the molecular radius. Rod-like larger molecules pass more easily than globular molecules.
2. Lipophilicity: Lipophilic substances usually diffuse more easily than lipophobic substances.
3. Ionization: Nonionized or weakly basic substances usually diffuse more easily than acidic substances.
4. pH: When the pH of the saliva in the sulcus is lower than that of the arterial blood in the capillary bed that is in contact with the acinar cells, basic substances are concentrated in the saliva relative to whole arterial blood.
5. Membrane transit: The free or unbound substance of interest usually is required to pass across the membranes into the initially formed saliva.

As shown in Fig. 1.2, sero-salivary substances also employ filtration through water-filled pores, active transport, or facilitated diffusion as a means to enter the initially formed saliva.

## DISEASE STATES THAT MAY INFLUENCE SALIVARY FUNCTION

Several disorders can have an adverse effect on salivation. Following is a partial list.

Alcoholic liver cirrhosis: In approximately 50% of patients, the parotid glands are enlarged, giving rise to a 50% reduction in salivary flow rate and a reduction of salivary sodium, bicarbonate, and chloride levels.[1]

Burning mouth syndrome: The term applies to several conditions of unknown etiology that are characterized by a burning sensation and pain. It is most often encountered in middle-aged women or people with candidiasis who have used antibiotics for an extended period of time.[1,9]

Cystic fibrosis (CF): CF is one of the most common autosomal recessive diseases in people of northern European ancestry.[10] It has been mapped to band 31.2 of the long arm of chromosome 7 (7q31.2). Because it is autosomal recessive, two aberrant genes must be present for it to be manifested phenotypically. Severity ranges from fatal in childhood to undetectable without unusual circumstances or targeted genotyping. When CF is present phenotypically, the oral fluid shows increased viscoelasticity with several electrolyte and protein concentration differences, compared to non-CF individuals.[1]

Graft-versus-host disease: This condition is precipitated by the immune response of histoincompatible, immunocompetent donor cells against tissues of an immunocompetent host. It may occur as a result of a complication of bone marrow transplantation or of maternal–fetal blood transfusion or a therapeutic blood transfusion in which the recipient has a cellular immunodeficiency disease.[9] Among the many manifestations, destruction of the salivary glands results in reduced salivary flow rate.[1]

Sjögren syndrome: This condition occurs usually in middle-aged or older women. It is complex and of unknown etiology. It includes keratoconjunctivitis sicca with or without lacrimal gland enlargement, xerostomia with or without salivary gland enlargement, and connective tissue disease which usually is rheumatoid arthritis but may be systemic lupus erythematosus, scleroderma, or polymyositis. An abnormal immune response may be implicated.[9]

Nerve damage: Since salivation is under the control of both the sympathetic and parasympathetic nervous systems, head and neck nerve damage may result in xerostomia.

Salivary gland removal: An obvious but rare cause of xerostomia is removal of the salivary glands either surgically or by cessation of function through exposure to radiation.

## THE INFLUENCE OF DRUGS ON SALIVARY FUNCTION

When oral fluid is collected for drug testing, the main purpose is the detection and, where appropriate, the quantification of drugs and their metabolites. However, not only may drugs have a significant effect on an individual's performance, they also may affect salivary function, possibly compromising oral fluid collection.

Any drug that interferes with the central nervous system or the peripheral nervous system will influence the production of saliva.[1]

Analgesics, antiarrhythmics, anticonvulsants, antidepressants, antiemetics, antihistamines, antihypertensives, anti-nausea agents, antiparkinson drugs, antipsychotics, antispasmodics, cytoxins, decongestants, diuretics, expectorants, monoamine oxidase inhibitors, and minor tranquilizers, which are all therapeutic drugs, will reduce salivation. Table 1.4 shows some therapeutic drugs that are anticholinergic (parasympathetic) and reduce salivary flow either as part of a deliberate anticholinergic action (left column) or as a side effect (right column).

Table 1.5 shows parasympathetic drugs that increase salivary flow rate.

| Table 1.4 Anticholinergic Therapeutic Drugs That Reduce Salivary Flow[1] | |
|---|---|
| **Anticholinergic Effect** | **Anticholinergic Side Effect** |
| Muscarine blockers<br>  atropine<br>  scopolamine<br>  clidinium<br>  ipratropium<br>  oxybutynin<br>  pirenzepine<br>  propantheline | Antidepressants<br>  amitriptyline<br>  desipramine<br>  imipramine<br>  lithium<br>  lofepramine<br>  maprotiline<br>  nortriptyline<br>  oxypropylene |
| Anti-Parkinson drugs<br>  benztropine<br>  biperiden<br>  orphenadrine<br>  trihexyphenidyl<br>  procyclidine | Antihistamines<br>  cyclizine<br>  diphenhydramine<br>  promethazine<br>  tripelennamine |
| | Antiarrhythmics<br>  disopyramide |
| | Antipsychotics<br>  Chlorpromazine<br>  Butyrophenones |

| Table 1.5 Parasympathetic Drugs That Increase Salivary Flow | |
|---|---|
| **Directly Acting** | **Indirectly Acting** |
| arecoline | cisapride |
| bethanechol | neostigmine |
| carbamylcholine | nizatidine |
| cevimeline | physostigmine |
| methacholine | |
| muscarine | |
| oxotremorine | |
| pilocarpine | |

Where appropriate to testing and interpreting results, references to drugs and disease states noted above will be cited.

## REFERENCES

1. Aps JKM, Martens LC. Review: The physiology of saliva and transfer of drugs into saliva. *Forensic Sci Int.* 2005;150:119−131.

2. Forde MD, Koka S, Eckert SE, Carr AB. Systemic assessments utilizing saliva. Part 1. General considerations and current assessments. *Int J Prosthodontics.* 2006;19:43−52.

3. Carpenter GH. The secretion components, and properties of saliva. *Annu Rev Food Sci Technol.* 2013;4:267−276.

4. Adeli K, Ceriotti F, Nieuwesteeg M. Reference information for the clinical Laboratory. In: Rifai N, Horvath AR, Wittwer CT, eds. *Tietz Textbook of Clinical Chemistry and Molecular Diagnostics.* 6th ed. Elsevier; 2018.

5. Höld KM, de Boer D, Zuidema J, Maes RAA. Saliva as an analytical tool in toxicology. *Int J Drug Test.* 2011;1−34.

6. Ritschel WA, Tompson GA. Monitoring of drug concentrations in saliva: a non-invasive pharmacokinetic procedure. *Meth Find Exptl Clin Pharmacol.* 1983;5(8):511−525.

7. SAMHSA, May 15, 2015 Federal Register, 80 FR 28053, Oral Fluid Mandatory Guidelines; Washington, DC, Government Printing Office.

8. Terrapon B, Mojon P, Mensi N, Cimasoni G. Salivary albumin of edentulous patients. *Arch Oral Biol.* 1996;41(12):1183−1185.

9. Dorland's Illustrated Medical Dictionary. 31st ed. Philadelphia, PA: Saunders Elsevier; 2007.

10. Vnencak-Jones CL, Best DH. Genetics. In: Rifai N, Horvath AR, Wittwer CT, eds. *Tietz Textbook of Clinical Chemistry and Molecular Diagnostics.* 6th ed. Elsevier; 2018.

# CHAPTER 2

# Oral Fluid Pharmacokinetics

## INTRODUCTION

Fluid generated in the oral cavity arises primarily from the major salivary glands, the minor salivary glands, the gingiva (crevicular fluid, minor contribution), and nasopharyngeal discharge, the contribution of which can vary depending on the state of health of the individual's nasal cavity (Fig. 2.1). Fluid from the lungs also may enter the oral cavity via the pharynx. Except under unusual circumstances, the oral cavity is fed primarily by the blood system, with minor contributions from the nasal cavity, pulmonary efflux, and the environment (e.g., smoke or dust). Except for occasional expectoration, essentially all the 500–1500 mL of oral fluid generated per day drains into the stomach via the esophagus. From the stomach, drained oral fluid is processed like any other liquid presented to the lower gastrointestinal tract. The oral cavity can be considered as a body compartment pharmacokinetically. This section will describe the basic pharmacokinetics of drugs of interest and their metabolites.

When a drug is administered intravenously, the drug and its metabolites may transfer from blood into the continuously forming oral fluid. For a drug or metabolite to do this, the barriers enumerated in Chapter 1 need to be overcome. To pass through those barriers, a drug or metabolite usually must be in the free or non-protein-bound form, nonionized, and of appropriate size. Rare exceptions may exist when a drug is bound to a protein that is excreted into saliva actively.

The pH of whole human blood is very tightly controlled by an extensive blood buffering system. Any deviations from the pH range stated in Table 1.3 generally result in clinically observable symptomatology. The table shows that the pH of human oral fluid is usually 6.2–7.4 or 5.6–7.9, depending on the reference. Under stress such as that induced by providing an oral fluid donor with sour candy or another salivary stimulant, the pH may rise as high as 8.0. The term "ion trapping" of a basic drug is many times quoted with the

Detection of Drugs and Their Metabolites in Oral Fluid. DOI: https://doi.org/10.1016/B978-0-12-814595-1.00002-7

**HUMAN DIGESTIVE SYSTEM**

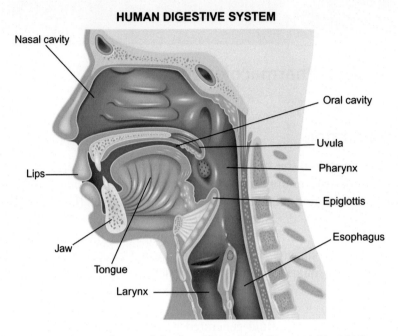

*Figure 2.1 Human oral cavity and associated physiological features.* From Shutterstock; Teguh Mujiono.

appearance of a drug from the blood in oral fluid. Ion trapping is where a basic drug demonstrates the usual protein unbound:bound equilibrium in whole blood (pH = approximately 7.35−7.45) but transfers readily into forming or formed saliva, which has a lower pH. The pH of oral fluid is usually, but not always, below that of the blood that feeds the salivary glands. Based solely on pH, one might intuitively conclude the following:

Basic Drug or Metabolite: Oral Fluid/Plasma > 1

Acidic Drug or Metabolite: Oral Fluid/Plasma < 1[1]

However, other factors such as lipophilicity and protein binding may enter into the equation. Additionally, not all drugs and their metabolites fit neatly into the acidic/basic/neutral categorization, as will be discussed under each drug or class below.

Please see the discussion below on effervescent fentanyl tablet for a commentary on the blood to saliva transfer system.

| Table 2.1 Average Oral Fluid-to-Blood Concentration Ratios for Selected Drugs | |
|---|---|
| Drug or Drug Class | Ratio |
| alcohol (ethanol) | 1.077 (1.094)[1] Please see "Ethanol" section, below |
| barbiturates | 0.3 |
| buprenorphine | 1 |
| codeine | 4 |
| methamphetamine | 2 |
| MDMA | 7 |
| cocaine | 3 |
| diazepam | 0.01–0.02 |
| methadone | 1.6 |
| morphine | 0.8 |
| $\Delta^9$-THC | 1.2 |

Average oral fluid-to-blood ratios of several drugs are presented in Table 2.1.[2] These numbers need to be used with the understanding that many factors, especially oral fluid pH, can influence the ratio.

The oral cavity is not a closed compartment. It is open not only at the mouth but also through the nostrils and nasal cavity and the esophagus. Although rarely taken into account, a drug that has been swallowed may reflux into the oral cavity and cause an unexpectedly elevated oral fluid drug level. Likewise, smoke (e.g., from a marijuana joint or a cigarette soaked in cocaine or methamphetamine) may contaminate the oral cavity and give rise to a markedly elevated drug level.

Insufflation ("snorting") is a common way to ingest cocaine in its salt form (usually hydrochloride). Insufflated drug that is not absorbed in the nasal passages may make its way into the oral cavity via nasopharyngeal fluids, resulting in an extremely elevated fluid concentration relative to a common dose.

Many drugs can be used by simple oral ingestion. This usually contaminates the oral cavity, producing an unexpectedly high oral drug level. The exact opposite can occur if a coated tablet is ingested without chewing or degradation of the coating. The drug in the coated tablet is not released until it arrives in the lower gastrointestinal tract, where it is absorbed; so the oral fluid level reflects only what is in the blood, assuming the drug will transfer from blood to oral fluid.

Following is a discussion of the fundamental classes of abused drugs of interest to toxicologists, with unusual circumstances kept in mind. Experience with the actual handling of oral fluid matrices and common drugs is reviewed in Chapter 9.

## ETHANOL

### Ethyl Alcohol, "Alcohol"

Where legal or at least readily available, ethanol (Scheme 2.1) is the most used and abused drug worldwide. For most clinical and forensic toxicology laboratories, determination of ethanol in blood and blood products, urine, and oral fluid is one of the most commonly performed analytical tests. The use of ethanol by humans has both acute and chronic effects, and it is a known teratogen. Acutely, ethanol is almost unique in that whole blood and blood product levels correlate extremely well with clinical symptoms.[3] Pharmacokinetically, ethanol's rate of disappearance is unusual in that it is zero order; i.e., a plot of a postabsorptive blood or blood product level ($y$-axis) against time ($x$-axis) yields a straight line.[3] Another almost unique property is that ethanol is a volatile liquid that is completely miscible with water.[4] In blood and blood products, it exists in the free form and is not bound to protein. It does not ionize at any pH compatible with life.[5] Its unique physiological and physical properties make it highly suitable as an oral fluid analyte both qualitatively and quantitatively.

Ethanol is taken into the human body almost exclusively by the oral route.

Perhaps due to the significantly greater amount of protein in plasma than in oral fluid (essentially saliva), the saliva-to-plasma ratio for ethanol has been stated to be 1.1.[1] As is true with other stated ratios for ethanol (e.g., serum-to-whole blood), the saliva-to-plasma ratio is a range rather than an absolute value. Jones stated it to be 1.077 with a range of 0.84−1.36 and later adjusted that to 1.094 with a range of

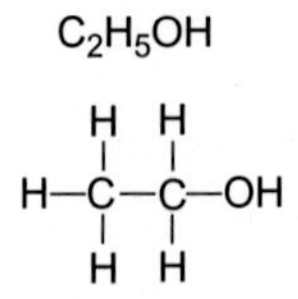

$$C_2H_5OH$$

$$H-\underset{\underset{H}{|}}{\overset{\overset{H}{|}}{C}}-\underset{\underset{H}{|}}{\overset{\overset{H}{|}}{C}}-OH$$

*Scheme 2.1 Structure of ethanol.*

0.88−1.36. Since a constant range exists for saliva-to-plasma, oral fluid ethanol can be used to demonstrate the correlation between oral fluid level and clinical symptoms. It is notable that before measuring a quantitative oral fluid ethanol level, a 20-min observed abstinence period should be used prior to sample collection, as is required for breath alcohol testing. As is true for blood and blood product ethanol elimination studies, the oral fluid donor must be postabsorptive to be in the linear phase.

## CANNABINOIDS

$\Delta^9$-Tetrahydrocannabinol (THC) (Scheme 2.2) is used as an antiemetic, analgesic, and appetite stimulant[4] as well as a recreational drug. It is most commonly smoked, but oral administration via so-called "edibles" appears to be gaining popularity, especially in areas where marijuana has been legalized. It also may enter oral fluid by passive exposure to smoke. Exposure to THC and other cannabinoids can be classified as active or passive, oral or smoked, and frequent or occasional, and the type of exposure can affect pharmacokinetics. The following paragraphs summarize current knowledge of oral fluid THC pharmacokinetics in each type of exposure briefly and within what is currently known at the time of this writing.

Although the target substance in urine drug testing usually is 11-nor-$\Delta^9$-tetrahydrocannabinol-9-carboxylic acid (THC-COOH, also called THCA), which is the primary metabolite of THC, it is the parent substance itself that is of primary concern in oral fluid drug testing. The single non-hydrocarbon moiety on THC is a phenolic group which is only weakly acidic ($pK_a = 10.6$). Parent THC is tightly bound to blood protein ($F_b = 0.97$).[6] Thus, THC is a poor candidate for transfer from blood to oral fluid. Consistent with parent THC's physical properties, contamination of the oral cavity with smoke mostly determines smoked THC oral fluid levels, rather than transfer of absorbed THC from blood to oral fluid.[7] The saliva-to-plasma ratio for parent THC has been calculated as 0.099−0.129[1] and is stated as 1.2 in a later reference.[2] At 0.060−0.099, the ratio of the active metabolite of THC, 11-hydroxy-$\Delta^9$-THC is even lower.[1]

Smoked marijuana in a study involving six healthy male subjects demonstrated the anticipated initial maximum for oral fluid THC at about 12 min. In corresponding plasma specimens, THC also peaked

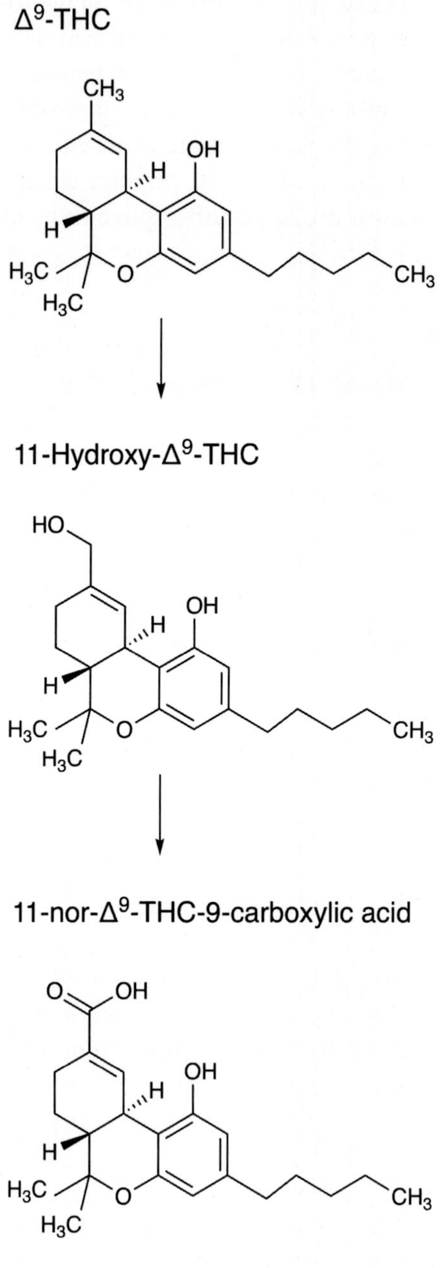

**Δ⁹-THC**

**11-Hydroxy-Δ⁹-THC**

**11-nor-Δ⁹-THC-9-carboxylic acid**

*Scheme 2.2 Metabolism of THC.*

at 12 min. A half-life ($t_{1/2}$) for THC in oral fluid was calculated to be 0.80 h, which was similar to the plasma half-life of 0.75 h. The authors suggested that transmucosal absorption of THC occurs during smoking and elevates existing blood levels.[8] In a separate study, the median $C_{max}$ for THC in oral fluid after smoking a cigarette containing 20 mg of THC (500 mg of cannabis with 4% THC) added to tobacco was 55.4−123,120.0 µg/L. The corresponding $C_{max}$ in plasma was 1.6−160.0 µg/L. It was stated that the variations in oral fluid ranges are too broad to predict plasma THC from the oral fluid values.[9] In a study that involved the use of the StatSure device for fluid collection, no statistically significant differences in the overall time course of the absorption, distribution, and elimination of cannabinoids was observed except for THC-COOH. The presence of THC-COOH identified fewer occasional smokers than THC. However, the presence of THC-COOH ruled out acute passive environmental exposure.[10]

In looking at cannabis edibles, Newmeyer et al.[11] generated the pharmacokinetic data shown in Table 2.2 following ingestion of approximately 50.6 mg of THC, 1.5 mg of cannabidiol, and 3.3 mg of cannabinol baked into a brownie. The brownie was consumed within 10 min. The participants were 10 frequent and 7 occasional marijuana smokers.

The results of Newmeyer et al.[11] reinforced the finding of Marsot et al.[9] that a reliable conversion cannot be obtained between oral fluid and blood cannabinoid values. Newmeyer et al. also noted that there was no oromucosal contamination by the drug dronabinol, because it was encapsulated. In a separate study by Vandrey et al.,[12] 18 participants ingested brownies containing 9.4, 23.6, or 48.5 mg (target values = 10, 25, and 50 mg) THC, and pharmacokinetic values were determined (Tables 2.3 and 2.4).

Although pharmacodynamics are beyond the scope of this section, the study by Vandrey et al. did include interesting cannabis pharmacodynamic outcomes. A previous paper that included the authors of the 2017 paper focused on the pharmacodynamics of cannabinoids from the second-hand smoke and effects of ventilation or lack thereof.[13]

Several studies on oral fluid cannabinoids resulting from passive inhalation have been done. Moore et al. found a maximum of 17 ng/mL of THC from exposure for 3 h in Dutch coffee shops. At a cutoff

## Table 2.2 Oral Fluid Pharmacokinetic Values for Cannabinoids and Metabolites After Brownie Consumption[11]

| Parameter | Frequent Smokers<br>Mean, Median (Range) | Occasional Smokers<br>Mean, Median (Range) | $t^a$ | $P^b$ |
|---|---|---|---|---|
| **THC** | | | | |
| $C_{max}$, µg/L | 573, 464 (39.3−2111) | 362, 392 (115−696) | 0.867 | 0.401 |
| $t_{max}$, h | 0.33 | 0.33 | | |
| $t_{last}$, h | 39, 44 (20−>48) | 23, 26 (20−26) | 3.876 | **0.003** |
| **11-OH-THC[c]** | | | | |
| $C_{max}$, µg/L | 0.6, 0.7 (0.2−1.2) | 0.4, 0.4 (0.3−0.6) | 1.521 | 0.156 |
| $t_{max}$, h | 0.40, 0.33 (0.33−1.0) | 0.60, 0.33 (0.33−1,5) | −0.983 | 0.347 |
| $t_{last}$, h | 1.1, 1.0 (0.33−3.5) | 1.4, 1.3 (0.33−3.5) | −0.333 | 0.745 |
| **THC-COOH** | | | | |
| $C_{max}$, ng/L | 329, 262 (123−1009) | 318, 191 (27.7−1281) | 0.062 | 0.951 |
| Adjusted $C_{max}$, ng/L | 285, 186 (123−849) | 315, 191 (27.9−1263) | −0.179 | 0.861 |
| $t_{max}$, h | 12, 5 (3.5−48) | 10, 10 (0.33−20) | 0.357 | 0.726 |
| $t_{last}$, h | >48 | 41, 44 (26−>48) | 1.943 | 0.100 |
| **THCV[d]** | | | | |
| $C_{max}$, µg/L | 7.4, 6.8 (1.3−19.4) | 5.4, 4.7 (1.6−10.6) | 0.800 | 0.436 |
| $t_{max}$, h | 0.33 | 0.33 | | |
| $t_{last}$, h | 2.4, 1.5 (1.5−3.5) | 1.9, 1.5 (1.0−3.5) | 0.852 | 0.409 |
| **CBD[e]** | | | | |
| $C_{max}$, µg/L | 14.0, 12.4 (2.0−29.7) | 10.8, 9.0 (4.0−18.4) | 0.807 | 00.433 |
| $t_{max}$, h | 0.33 | 0.33 | | |
| $t_{last}$, h | 3.0, 3.5 (1.5−5.0) | 2.6, 3.5 (1.0−3.5) | 0.708 | 0.491 |
| **CBG[f]** | | | | |
| $C_{max}$, µg/L | 31.2, 27.2 (3.5−90.1) | 21.2, 22.6 (7.5−33.9) | 0.936 | 0.365 |
| $t_{max}$, h | 0.33 | 0.33 | | |
| $t_{last}$, h | 3.6, 3.5 (1.5−5.0) | 4.6, 3.5 (1.0−14) | −0.643 | 0.531 |

[a]*Student's t-test.*
[b]*Bold P-value indicates significance.*
[c]*11-hydroxytetrahydrocannabinol.*
[d]*Tetrahydrocannabivarin.*
[e]*Cannabidiol.*
[f]*Cannabigerol.*

of $\geq 20$ pg/mL, no THC-COOH was found, indicating that an absence of THC-COOH might be useful in some cases to show that THC in oral fluid was due to passive inhalation and not use.[14] In a study by Niedbala et al., it was noted that the oral fluid collection device employed in passive inhalation studies must be stored and used outside the area where it might be exposed to marijuana smoke.[15]

**Table 2.3 Oral Fluid Concentrations of THC and THC-COOH After Brownie Consumption (Vandrey et al.)**

| Target Dose of THC (mg) | THC | THC | THC-COOH | THC-COOH |
|---|---|---|---|---|
| | $C_{max}$ (ng/mL) | $t_{max}$ (h) | $C_{max}$ (ng/mL) | $t_{max}$ (h) |
| 10 | 191.5 (47−412) | 0.2 (0.2−0.5) | 0.051 (0−0.231) | 1.0 (0−3) |
| 25 | 477.5 (70−1128) | 0.2 (0.2−0.5) | 0.140 (0.023−0.251) | 9.8 (3−30) |
| 50 | 597.5 (350−1010) | 0.2 (0.2−0.5) | 0.314 (0−0.822) | 17.4 (0−54) |

*11-Hydroxy-THC $C_{max}$ and $t_{max}$ were not determined in oral fluid.*

**Table 2.4 Detection Time to First and Last Positives in the Brownie Study by Vandrey et al.**

| | Detection Time to First Positive (h) | | | |
|---|---|---|---|---|
| Target Dose of THC (mg) | THC | THC-COOH | THC | THC-COOH |
| | ELISA | ELISA | LC-MS/MS | LC-MS/MS |
| | (Cutoff = 4 ng/mL) | (Cutoff = 4 ng/mL) | (h, LOQ = 1 ng/mL) | (Cutoff = 0.050 ng/mL) |
| 10 | 0.17 (0.2−0.2) | 0.17 (0.2−0.2) | 0.17 (0.2−0.2) | 3.0 (3.0−30) |
| 25 | 0.23 (0.2−0.5) | 0.23 (0.2−0.5) | 0.23 (0.2−0.5) | 3.6 (0.2−8.0) |
| 50 | 0.17 (0.2−0.2) | 0.17 (0.2−0.2) | 0.17 (0.2−0.2) | 1.6 (0.2−4.0) |
| | **Detection Time to Last Positive (h)** | | | |
| 10 | 1.5 (1,2) | 0.7 (0−2) | 1.9 (2,3) | 1.0 (0−3) |
| 25 | 2.3 (1−4) | 16.4 (1−50) | 3.0 (2−6) | 26.3 (0−70) |
| 50 | 10.0 (2−22) | 25.2 (3−70) | 9.5 (3−22) | 37.3 (0−78) |

THC's major metabolite THC-COOH, which is inactive, usually can be found in oral fluid but at only about 1-1000th of the concentration of the parent.[16] It is formed by the action of cytochrome P4502C9 (CYP2C9),[17] which produces the active 11-hydroxy-THC, which is further oxidized to the inactive THC-COOH by a series of hydroxylases. It is found in oral fluid as about two-third free and one-third glucuronide conjugate.[18] The glucuronide is formed by uridine diphosphate glucuronosyltransferase 1A3 and 1A1.[19] THC-COOH in oral fluid is presumed to be formed from THC presented to the human liver. It reenters the bloodstream and is transferred to saliva. However, at the time of this writing, the formation of THC-COOH in oral cavity cells that are bathed with oral fluid containing THC had not been ruled out.[20−22] In one controlled smoking study of 10 subjects, the maximum THC-COOH concentration occurred 1−2 h after the start of smoking except for two participants in whom it occurred at 0.25 h.[22]

## COCAINE

Parent cocaine (Scheme 2.3), which is the active principal, can be insufflated, injected intravenously, smoked, and taken orally. The oral route is wasteful due to the amount hydrolyzed in stomach acid. Parent cocaine is basic ($pK_a = 8.6$), with a blood protein binding of only slightly less ($F_b = 0.92$)[23] than THC. The half-life in blood is approximately 0.7–1.5 h. The saliva terminal half-life has been reported as 7.9 h.[1]

Not unexpectedly for a basic drug, the saliva-to-plasma ratio shifts from 5 or greater in unstimulated saliva down to 0.5–3 in stimulated saliva.[1]

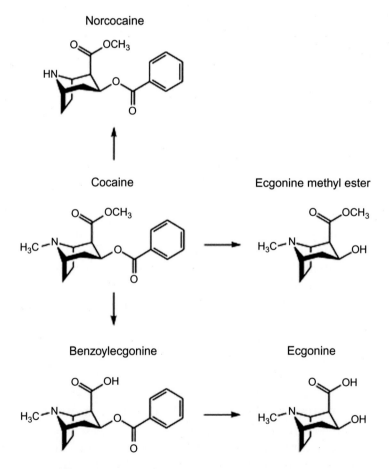

*Scheme 2.3 Metabolism of cocaine.*

Unlike THC, when cocaine is administered intravenously, it appears rapidly in oral fluid. In a study of 10 cocaine users administered 25 mg alone and in combination with oral acetazolamide and oral quinine, the pharmacokinetic parameters for cocaine and its major metabolite benzoylecgonine (BZE) were measured. Tables 2.5 and 2.6[24] show the results when cocaine was administered alone.

Regardless of the type of oral fluid collector, the half-life for cocaine in oral fluid was found to be similar to that for blood plasma.

Please see Chapter 7 for a discussion of the effect of oral fluid collection device on drug detection times.

## AMPHETAMINES

Amphetamine, methamphetamine, and methylenedioxyamphetamine (MDA) are more basic than cocaine (Table 2.7). Methylenedioxymethamphetamine (MDMA, ecstasy) is about as basic as cocaine, but its protein binding is intermediate between cocaine on the high side and amphetamine and $d$-methamphetamine on the low

| Table 2.5 Oral Fluid Pharmacokinetics for Cocaine and Its Major Metabolite Using the Oral-Eze Collector (Range in Parentheses) | | |
|---|---|---|
| Parameter | Cocaine | BZE |
| $C_{max}$ ($\mu$g/L) | 932 (394–1574) | 248 (96.9–953) |
| $t_{max}$ (h) | 0.34 (0.17) | 0.5 (0.17–1) |
| $t_{1/2}$ (h) | 1.3 (0.6–2.2) | 6.9 (4.4–12.1) |
| $t_{last}$ (h) | 12.5 (4–69) | 30.5 (21–69) |
| $C_{last}$ ($\mu$g/L) | 1.9 (1.0–35.1) | 2.8 (1.1–20.8) |

| Table 2.6 Oral Fluid Pharmacokinetics of Cocaine and Its Major Metabolite Using the StatSure Collector (Range in Parentheses) | | |
|---|---|---|
| Parameter | Cocaine | BZE |
| $C_{max}$ ($\mu$g/L) | 732 (83.3–1892) | 360 (77.2–836) |
| $t_{max}$ (h) | 0.17 (0.17–0.50) | 0.5 (0.17–1.50) |
| $t_{1/2}$ (h) | 0.89 (0.57–1.4) | 6.6 (4.1–14.1) |
| $t_{last}$ (h) | 6.5 (4–45) | 28 (21–69) |
| $C_{last}$ ($\mu$g/L) | 5.3 (1.0–26.6) | 3.5 (1–11.4) |

**Table 2.7 Pharmacokinetic Properties of Cocaine and Amphetamines**[23,25−29]

| Drug | pK$_a$ | $F_b$ | $V_d$ (L/kg) | Plasma $t_{1/2}$ (h) |
|------|------|------|------|------|
| cocaine | 8.6 | 0.92 | 1.6−2.7 | 0.7−1.5 |
| amphetamine | 9.9 | 0.16 | 3.2−5.6 | 7−34 (urine pH-dependent) |
| d-methamphetamine | 9.9 | 0.10−0.20 | 3.0−7.0 | 6−15 (urine pH-dependent) |
| l-methamphetamine | 9.9 | ? | 3−5 | 11−16 |
| MDA | 9.7 | ? | ? | ? |
| MDMA | 8.7 | 0.65 | 3−7 | 5−9 |
| MDMA, methylenedioxymethamphetamine ("ecstasy"). | | | | |

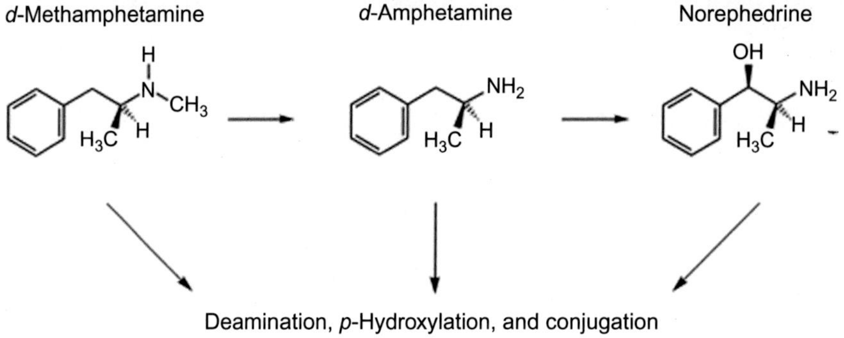

Scheme 2.4 Metabolism of methamphetamine.

side. Thus amphetamine, both methamphetamine isomers, MDA, and MDMA are all transferred easily from blood to saliva.

## Methamphetamine

Methamphetamine can be smoked, but most amphetamines are ingested (Scheme 2.4).

The oral fluid-to-blood ratios of amphetamine, methamphetamine, and MDMA have been estimated as 7.4, 4.5, and 5.1, respectively, using the first-generation Intercept collection device.[30]

In a study by Schepers et al., four oral 10-mg sustained-release d-methamphetamine doses were administered to eight participants within 7 days. Three weeks later, five participants received four oral 20-mg doses. Oral fluid was collected for up to 72 h after each administration. Blood was collected for up to 24 h. A citric acid sourball candy was used to stimulate oral fluid production in the first two sessions. In the third session, oral fluid was collected by putting a cotton swab treated

**Table 2.8 Oral Fluid Pharmacokinetics for $d$-Methamphetamine in the Study by Schepers et al.[31]**

| Parameter | 10-mg Methamphetamine Dose | | 20-mg Methamphetamine Dose | |
|---|---|---|---|---|
| | Mean ± SD | Range | Mean ± SD | Range |
| $t_{max}$ (h) | 5.0 ± 1.9 | 4.0−8.0 | 4.7 ± 3.9 | 2.0−11.5 |
| $C_{max}$ (μg/L) | 106.1 ± 100.8 | 24.7−312.2 | 192.2 ± 120.8 | 75.3−321.7 |
| $t_{1/2}$ (h) | 7.1 ± 2.3 | 3.2−10.9 | 8.1 ± 1.9 | |
| $V_d$ (L/kg) | 1.7 ± 0.9 | 0.6−3.2 | 2.6 ± 2.1 | 5.5−10.5 |

**Table 2.9 $t_{max}$ and $C_{max}$ for the Metabolite Amphetamine After $d$-Methamphetamine in the Study by Schepers et al.[31]**

| Parameter | 10-mg Methamphetamine Dose | | 20-mg Methamphetamine Dose | |
|---|---|---|---|---|
| | Mean ± SD | Range | Mean ± SD | Range |
| $t_{max}$ (h) | 9.1 ± 3.0 | 4.0−12.0 | 8.2 ± 3.5 | 2.0−12.0 |
| $C_{max}$ (μg/L) | 8.6 ± 6.5 | 3.8−21.3 | 14.3 ± 6.1 | 2.8−20.2 |

with 20 mg of citric acid into the donor's mouth. In the fourth session, cotton swabs without citric acid were used.[31] Tables 2.8 and 2.9 show the pharmacokinetic results.

Table 2.9 shows the results for the metabolite amphetamine.

It is notable that the half-life for methamphetamine in oral fluid is similar to that in plasma, although it tends toward the low side of the plasma range. The volume of distribution ($V_d$) is decidedly lower than in plasma but still on the same order of magnitude. The peak concentration of amphetamine in the study by Schepers et al. is less than 10% of that for methamphetamine. Not unexpectedly, the $t_{max}$ for amphetamine is slightly longer than for methamphetamine.

In the study by Schepers et al., the mean pH of oral fluid specimens harvested with the citric acid−treated swabs was on average 1.5 pH units lower than those collected after citric acid candy (swab 2.8 ± 0.3, candy 4.3 ± 0.8). The candy stimulation yielded values approximately 1.7 pH units lower than neutral cotton swabs (6.0 ± 0.6). The data suggested that the disposition of methamphetamine in oral fluid may be dose-related. However, high intrasubject and intersubject variability restrict the use of a single oral fluid measurement to predict a simultaneously obtained plasma concentration.

**Table 2.10 $C_{max}$, $t_{max}$, and Time to First Positive at LOQ for Methamphetamine and its Metabolite Amphetamine**

| Methamphetamine | | | |
|---|---|---|---|
| Dose of Methamphetamine (mg) | $C_{max}$ (ng/mL) | $t_{max}$ (h) | Time to First Positive (h) |
| 10 | 57.1 ± 12.9 | 5.6 ± 1.0 | 1.6 ± 0.2 |
| 20 | 192 ± 54.0 | 4.7 ± 1.8 | 0.8 ± 0.1 |
| Amphetamine | | | |
| 10 | 6.5 ± 1.5 | 10.3 ± 1.2 | 6.5 ± 2.5 |
| 20 | 14.3 ± 3.0 | 8.2 ± 1.7 | 4.1 ± 1.9 |

In another study of methamphetamine disposition, Huestis and Cone looked at $C_{max}$, $t_{max}$, and time to first positive at a stated LOQ (limit of quantification or quantitation) for methamphetamine and its metabolite amphetamine[32] (Table 2.10). With increasing methamphetamine dose, the $C_{max}$ for both methamphetamine and amphetamine increased but was higher than anticipated for the higher dose. Likewise, the $t_{max}$ and time to first positive shifted to lower values for both the drug and its metabolite. It was noted that urine testing usually gives high detection rates for methamphetamine and its metabolite for up to 3 days following drug cessation. In contrast, oral fluid testing yields only a moderate detection rate for 24 h.

Please see Chapter 7 for further presentation on chirality of amphetamine and methamphetamine.

## Methylenedioxymethamphetamine
Barnes et al. studied MDMA and its metabolite MDA (Scheme 2.5), and included monitoring of 4-hydroxy-3-methoxymethamphetamine, and 4-hydroxy-3-amphetamine (Table 2.11).[33] The authors concluded that oral fluid is a good alternate matrix for monitoring MDMA and can detect a single recreational dose of 70−150 mg for up to 1 or 2 days after use. It is notable that the half-life of MDMA in oral fluid is dose-related and very similar to its half-life in plasma (Table 2.11).

**Table 2.11 Oral Fluid Pharmacokinetic Mean Parameters for MDMA and its Metabolite MDA**

| | | MDMA | | |
|---|---|---|---|---|
| Dose of MDMA (mg/kg) | Last Detected (h, at LOQ = 5 ng/mL) | $C_{max}$ (ng/mL) | $t_{max}$ (h) | $t_{1/2}$ (h) |
| 1.0 | 36.5 (29.0−74.0) | 1643.0 (1160.0−3382.0 | 2.8 (1.3−5.0) | 4.6 (3.2−11.4) |
| 1.6 | 47.0 (47.0−71.0) | 4760.0 (2881.0−11,985.0) | 2.6 (1.5−4.5) | 7.4 (5.9−13.4) |
| | | MDA | | |
| 1.0 | 23.0 (23.0−34.0) | 41.0 (23.0−151.0) | 4.8 (2.8−23.0) | 8.8 (4.6−54.0) |
| 1.6 | 39.0 (29.0−47.0) | 128.0 (50.0−403.0) | 4.5 2.5−15.0) | 8.1 (6.9−22.8) |

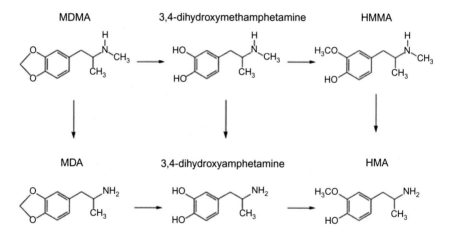

*Scheme 2.5 Metabolism of MDMA.*

## OPIATES

### Codeine, Morphine, and Diacetylmorphine

Among the many naturally derived opiates and their synthetic derivatives (Scheme 2.6), codeine, morphine, hydrocodone, hydromorphone, oxycodone, oxymorphone, and diacetylmorphine (heroin) stand out as being most common. Except for heroin, which can be smoked, snorted, or injected, most opiates are ingested. Basicity of the common opiates is moderate and protein binding is low to moderate, as can be seen in Table 2.12.[34−40] Thus, opiates usually are transferred fairly easily from whole blood to saliva.

## Table 2.12 Oral Fluid pK$_a$ and Fraction of Protein Bound for Common Opiates

| Drug | pK$_a$ | $F_b$ |
|---|---|---|
| codeine | 8.2 | 0.07–0.25 |
| morphine | 7.9 | 0.35 |
| hydrocodone | 8.9 | 0.25 |
| hydromorphone | 8.2 | 0.19 |
| oxycodone | 8.5 | 0.45 |
| oxymorphone | 8.5 | 0.10–0.12 |
| diacetylmorphine (heroin) | 7.6 | <0.05 |

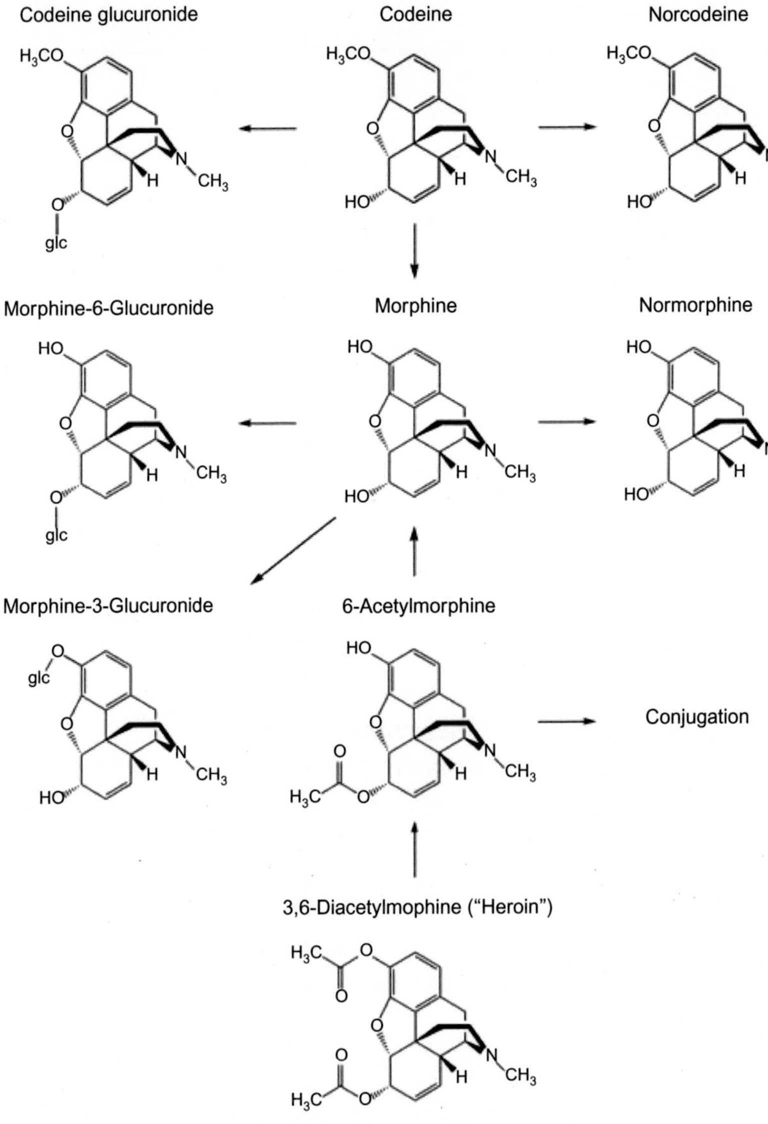

*Scheme 2.6 Metabolism of codeine, morphine, and diacetylmorphine.*

Using 12 male and 7 female subjects and dosing at two levels (60 and 120 mg/70 kg), Kim et al. generated the pharmacokinetic data in Table 2.13 for codeine and its metabolite norcodeine.[41] The authors noted that norcodeine is less basic and more polar than codeine, possibly resulting in considerably lower amounts of norcodeine than codeine in oral fluid.

The corresponding half-lives of codeine in plasma, $2.1 \pm 0.08$ h and $2.4 \pm 0.18$ h, are somewhat longer but similar to those in oral fluid. Likewise, the half-lives for norcodeine in plasma, $4.3 \pm 0.61$ h and $4.3 \pm 0.45$ h, are approximately twice the half-lives in oral fluid, but similar.

## Hydrocodone
Cone et al. generated data in 12 subjects who received a total of 20 mg of hydrocodone bitartrate (12.1 mg of free base; Table 2.14).[42] Hydromorphone was not seen in any of the oral fluid samples (Scheme 2.7).

**Table 2.13 Oral Fluid Pharmacokinetic Values for Codeine and its Metabolite Norcodeine**

| Codeine | | | |
|---|---|---|---|
| Dose of Codeine | $C_{max}$ ($\mu$g/L) | $t_{max}$ (h) | $t_{1/2}$ (h) |
| 60 | $638.6 \pm 64.4$ | $1.7 \pm 0.3$ | $2.5 \pm 0.21$ |
| 120 | $1599.3 \pm 241.0$ | $1.6 \pm 0.14$ | $1.8 \pm 0.19$ |
| Norcodeine | | | |
| 60 | $17.1 \pm 3.1$ | $2.1 \pm 0.31$ | $7.9 \pm 2.17$ |
| 120 | $46.7 \pm 17.0$ | $2.4 \pm 1.19$ | $4.6 \pm 1.11$ |

**Table 2.14 Oral Fluid Pharmacokinetic Values for Hydrocodone and its Metabolites Norhydrocodone and Dihydrocodeine**

| Parameter | Hydrocodone | Norhydrocodone | Dihydrocodeine |
|---|---|---|---|
| $C_{max}$ (ng/mL) | $207.7 \pm 42.0$ (61.7–625.6) | $12.8 \pm 2.4$ (3.6–27.0) | $6.4 \pm 1.2$ (2.6–18.2) |
| $t_{max}$ (h) | $1.4 \pm 0.3$ (0.3–4.0) | $2.6 \pm 0.6$ (1.0–8.0) | $4.5 \pm 0.9$ (1.5–12.0) |
| $t_{1/2}$ (h) | $4.4 \pm 0.2$ (3.2–5.6) | $6.2 \pm 0.7$ (3.4–10.2) | $5.7 \pm 0.9$ (2.6–10.2) |
| Mean Residence Time (h) | $6.2 \pm 0.4$ (4.3–8.4) | $11.3 \pm 1.9$ (6.4–25.0) | $10.2 \pm 0.9$ (7.0–14.3) |

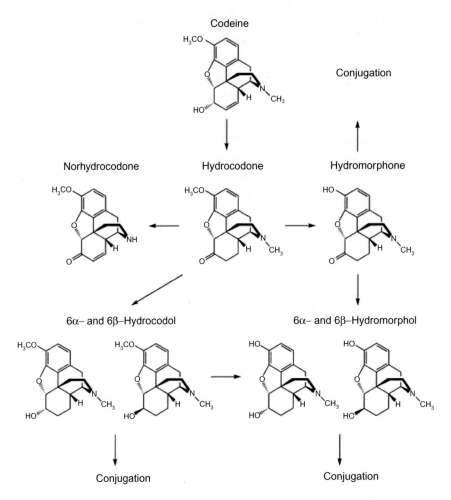

*Scheme 2.7 Metabolism of hydrocodone.*

The plasma half-life of hydrocodone is given as 3.4—8.8 h. Like codeine, the half-life of hydrocodone in plasma is similar but longer than in oral fluid.[36]

## Oxycodone

Data from another study in which 12 subjects received one 20-mg oxycodone controlled release tablet (Oxycontin) are presented in Table 2.15 (Scheme 2.8).[43]

**Table 2.15 Oral Fluid Pharmacokinetic Values for Controlled Release Oxycodone and its Metabolites Noroxycodone and Oxymorphone**

| Parameter | Oxycodone | Noroxycodone | Oxymorphone |
|---|---|---|---|
| $C_{max}$ (ng/mL) | 132.7 ± 15.4 (49.2−218.7) | 18.7 ± 1.6 (10.3−31.8) | 1.6 ± 0.2 (1.2−2.4) |
| $t_{max}$ (h) | 3.3 ± 0.4 (2.0−6.0) | 5.1 ± 0.6 (2.5−8.0) | 3.7 ± 0.3 (2.0−4.0) |
| $t_{1/2}$ (h) | 4.6 + 0.3 (3.4−5.9) | 8.3 ± 0.8 (3.6−13.3) | NA |
| Mean Residence Time (h) | 9.7 ± 0.5 (7.5−14.0) | 14.4 ± 0.8 (10.7−19.9) | NA |
| *In all cases, noroxymorphone was noted as NA (not applicable).* | | | |

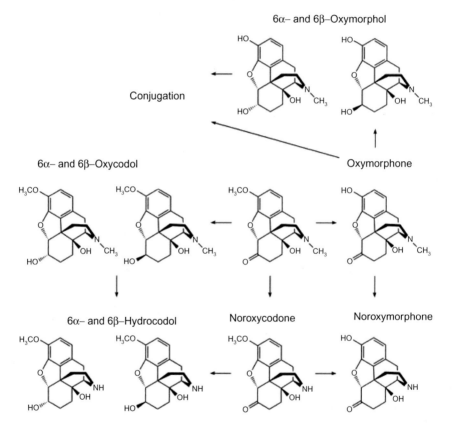

Scheme 2.8 Metabolism of oxycodone.

The half-life of oxycodone in plasma is listed as 3−6 h,[38] similar to its half-life in oral fluid.

## BENZODIAZEPINES

Benzodiazepines (Table 2.16) are strongly bound to plasma proteins and have $pK_a$ values usually much less than 6−7. Thus, they tend to

| Table 2.16 Oral Fluid pK$_a$ and Fraction of Protein Bound for Common Benzodiazepines[44-48] | | |
|---|---|---|
| Drug | pK$_a$ | F$_b$ |
| alprazolam | 2.4 | 0.65−0.75 |
| diazepam | 3.4 | 0.96 |
| flunitrazepam | 1.8 | 0.78 |
| flurazepam | 1.9, 8.2 | 0.97 |
| triazolam | 1.2, 1.5, 2.2, 6.5 | 0.78 |

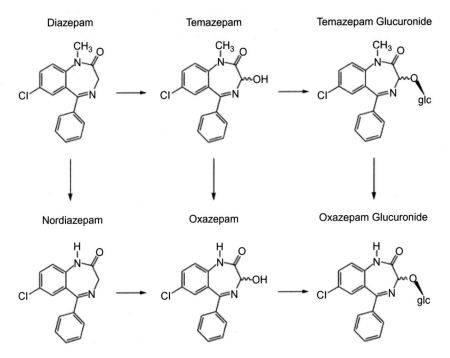

Scheme 2.9 Metabolism of diazepam.

transfer poorly from plasma to saliva. Saliva-to-plasma ratios have been reported as 0.01−0.08, imposing a requirement for extremely sensitive methods to monitor them in oral fluid.[1]

The majority of benzodiazepines currently marketed follow either the diazepam pathway (Scheme 2.9) or the alprazolam pathway (Scheme 2.10).

Please see Chapter 7 for a discussion on the recovery of parent nitrobenzodiazepines versus their amino metabolites.

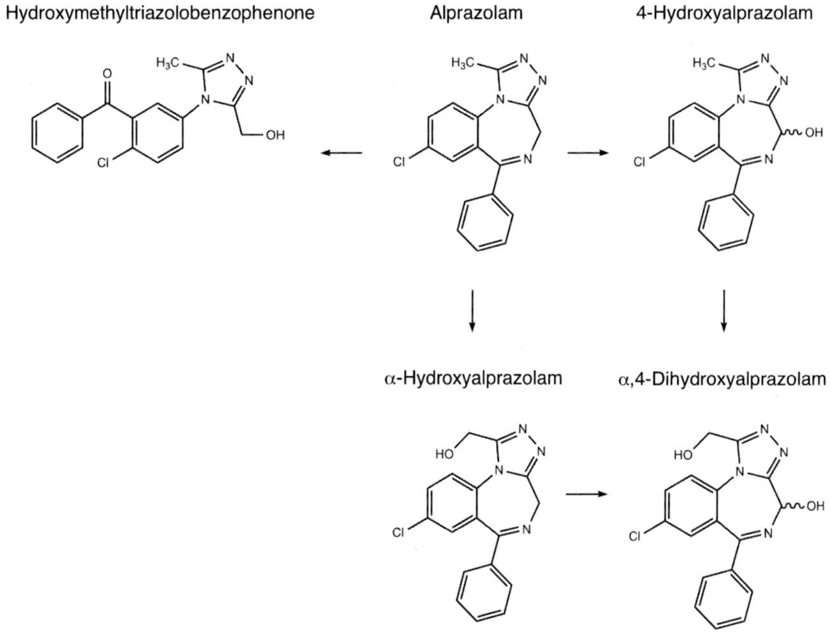

Hydroxymethyltriazolobenzophenone          Alprazolam          4-Hydroxyalprazolam

α-Hydroxyalprazolam      α,4-Dihydroxyalprazolam

*Scheme 2.10 Metabolism of alprazolam.*

## OTHER DRUGS

### Phencyclidine

Although a true positive (not a blind control) sample for phencyclidine is seen only rarely, it is included in the Federal Guidelines for employment-related urine drug testing. Thus, it is worth mentioning that because of its low plasma protein binding ($F_b < 10\%$) and $pK_a$ of 9.43, it can be identified readily in oral fluid.[1]

### Methadone

Methadone (Scheme 2.11) and its metabolite 3-ethylidine-1,5-dimethyl-5,5-diphenylpyrrolidine (EDDP) can be monitored in oral fluid. With a $pK_a$ of 8.2, the saliva-to-plasma ratio is a function of salivary pH. The ratio of the parent drug has been reported as 0.6–7.2, varying as a function of salivary pH. For EDDP in the same specimens, the ratio varied from 0.2 to 1.8 and was not dependent on pH.[1]

### Fentanyl

Fentanyl is a powerful opioid analgesic.[49] It is normally delivered transdermally,      subcutaneously,      intravenously,      intramuscularly,

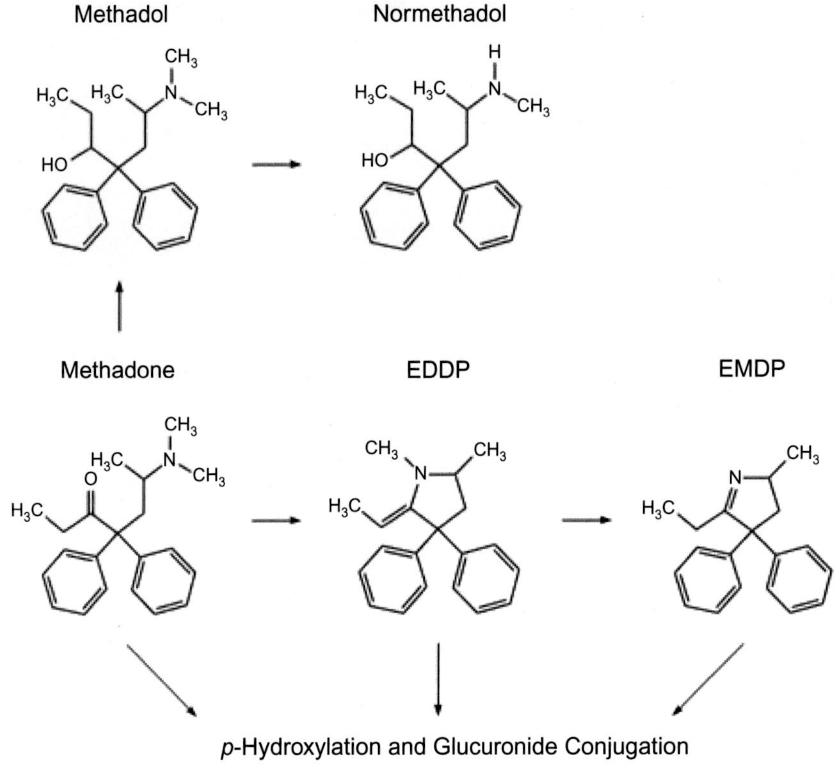

Scheme 2.11 Metabolism of methadone.

transmucosally, and by spinal injection.[50] Smoking and ingestion also are possible for illicit use. Basic metabolic conversion is accomplished mostly by CYP3A4 (Scheme 2.12).

Table 2.17 shows pharmacokinetic parameters for fentanyl delivered transmucosally from an effervescent buccal tablet.[51]

The transmucosal drug delivery system,[51] which uses fentanyl cit-rate in a carbon dioxide-generating tablet, also provides an opportu-nity to view the system of transfer from oral fluid-to-blood, which is the reverse of the system in which drugs and their metabolites are transferred from blood to saliva, which becomes oral fluid. The effer-vescence produces carbon dioxide which results in a dynamic shift in pH as the tablet dissolves. The resulting low pH favors dissolution of the fentanyl citrate in oral fluid. As the carbon dioxide dissipates, the pH increases, favoring buccal absorption of nonionized fentanyl across the buccal mucosa. The "pH pumping" mechanism increases the

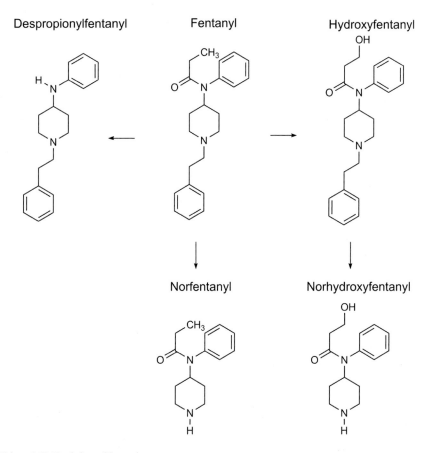

*Scheme 2.12 Metabolism of fentanyl.*

**Table 2.17 Pharmacokinetic Values for Fentanyl Delivered Transmucosally From an Effervescent Buccal Tablet**

| Dose (μg) | $t_{max}$ (h) (median, 90% CI) | $C_{max}$ (ng/mL) (mean ± SD) | $t_{1/2}$ (h) (median, 90% CI) |
|---|---|---|---|
| 100 | 0.75 (0.4–3.02) | 0.25 ± 0.14 | 2.63 (1.5–13.6) |
| 200 | 0.67 (0.33–3.0) | 0.40 ± 0.18 | 4.43 (1.9–20.8) |
| 400 | 0.58 (0.33–3.0) | 0.97 ± 0.53 | 11.1 (3.4–20.6) |
| 800 | 0.67 (0.4–3.0) | 1.59 ± 0.90 | 11.7 (4.6–28.6) |
| *CI, confidence interval.* | | | |

permeation of fentanyl into and through the buccal membrane into the vascular system, where the agent is transported to a specific opioid receptor site. The reverse of the transmucosal drug delivery system minus the in situ generation of carbon dioxide is the mechanism by which basic drugs are transported into saliva and trapped there.

**Table 2.18 Blood Pharmacodynamic and Pharmacokinetic Parameters of Fentanyl and Analogues[49,54−57]**

| Drug | $F_b$ | pKa | $V_d$ (L/kg) | $t_{1/2}$ | Potency (× morphine) |
|------|-------|-----|--------------|-----------|----------------------|
| fentanyl | 0.79 | 8.4 | 3−8 | 3−12 h | 100 |
| alfentanil | 0.92 | 6.5 | 0.3−1.0 | 1−2 h | 10−25 |
| carfentanil | Not approved for human use | | | | 9000 |
| remifentanil | 0.70 | 7.1 | 0.2−0.4 | 6−16 min | 420−1675 |
| sufentanil | 0.90 | 8.0 | 0.93 | 1.6−5.7 h | 500−800 |

Although fentanyl exists as a separate drug, there are several hundred variants. In one publication on overdose deaths due to fentanyl and its analogues in Ohio, the cited methodology tested for 25 fentanyl analogues, metabolites, and structurally unrelated synthetic opioids.[52] Another publication cites overdoses from ortho-fluorofentanyl, which is the result of modifying the original structure with a halogen,[53] The blood pharmacokinetic and potency (pharmacodynamics) parameters for fentanyl and several common analogues are presented in Table 2.18.

The potency of fentanyl and its analogues should provide a guideline for the levels that may be encountered in oral fluid. It is worth noting that the half-life of fentanyl is similar in plasma and oral fluid.

Even though oral fluid and plasma fentanyl levels appear to be correlated with dose, oral fluid levels usually are significantly higher.[58−60] Norfentanyl in oral fluid is present in sufficient quantity to make it useful for test results interpretation.[61] Salivary hypofunction has been shown to have a negative effect on the absorption of a sublingual citrate formulation.[62]

## Nicotine

Nicotine is easily available in most areas of the world through cigars, cigarettes, chewing tobacco, and other tobacco products. Because it is so common, it is often overlooked by toxicologists. However, the presence of nicotine or its metabolites can have serious implications when obtaining life and health insurance or in assessing compliance with smoking cessation. Nicotine and eight of its metabolites or biomarkers found with nicotine (nicotine-N-β-D-glucuronide, cotinine-N-oxide, *trans*-3-hydroxycotinine, norcotinine, nornicotine, anatabine, anabasine, and cotinine-N-β-D-glucuronide) can be monitored in human oral fluid.[63] The oral fluid-to-plasma ratio for nicotine was found to be 6.4 and for cotinine 3.3 (Scheme 2.13).

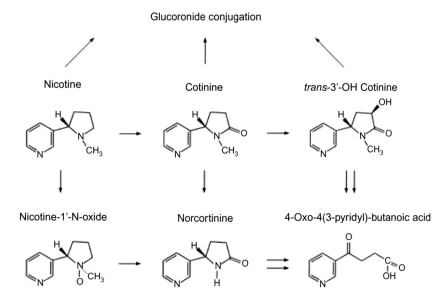

Scheme 2.13 Metabolism of nicotine.

## NEW PSYCHOACTIVE SUBSTANCES

New psychoactive substances (NPS) is a broad category that includes synthetic cannabinoids (cannabimimetics), synthetic cathinones, phenethylamines, tryptamines, arylalkylamines, synthetic cocaines, opioids and opioid analogues, benzodiazepines, and other compounds that mimic current known drugs.[64] Due to the size of the category, targeting by approximate class (e.g., benzodiazepines, fentanyl analogues, and phenethylamines) is encouraged, although many times a common drug such as MDMA or MDA may be detected in a procedure designed primarily for cathinones.[65]

Although several methods have been published for NPS,[66–68] flexibility is key to results production due to the rapid rate with which compounds and their metabolites appear on the market and disappear as saleable products. Individual NPS may become common and then disappear in a matter of months due to a number of circumstances, including but not limited to the disappearance of the chemist who manufactured the original batch, a limited supply of precursor, and synthesis of a different analogue such as halogenation of a specific site on the original compound.[69] Because of this flux, a complete study of each new substance is difficult if not impossible for researchers. Thus, a laboratory must be careful with reporting to indicate to the client whether only parent substances are included in an oral fluid report and if any metabolism is known or not.

# REFERENCES

1. Spiehler V. Drugs in Saliva. In: Moffat AC, Osselton MD, Widdop B, eds. *Clarke's Analysis of Drugs and Poisons.* 3rd ed. London: Pharmaceutical Press; 2004.

2. Drummer OH. Drug testing in oral fluid. *Clin Biochem Rev.* 2006;27:147−159.

3. Porter WH. "Clinical toxicology". In: Burtis CA, Ashwood ER, Bruns DE, eds. *Tietz Textbook of Clinical Chemistry and Molecular Diagnostics.* 4th ed. Saint Louis: Elsevier Saunders; 2006.

4. The Merck Index. 15th ed. Cambridge: The Royal Society of Chemistry; 2013.

5. Baselt RC, ed. *Disposition of Toxic Drugs and Chemicals in Man.* 10th ed. Seal Beach, CA: Seal Beach, Biomedical Publications; 2014, p. 781.

6. Baselt RC, ed. *Disposition of Toxic Drugs and Chemicals in Man.* 10th ed. Seal Beach, CA: Seal Beach, Biomedical Publications; 2014, p. 1948.

7. Lee D, Huestis MA. Current knowledge on cannabinoids in oral fluid. *Drug Test Anal.* 2013;6(2):88−111.

8. Huestis MA, Cone EJ. Relationship of $\Delta^9$-tetrahydrocannabinol concentrations in oral fluid and plasma after controlled administration of smoked cannabis. *J Anal Toxicol.* 2004;28 (6):394−399.

9. Marsot A, Audebert C, Attolini L, Lacarelle J, Blin O. Comparison of cannabinoid concentrations in plasma, oral fluid and urine in occasional cannabis smokers after smoking cannabis cigarette. *J Pharm Pharm Sci.* 2016;19(3):411−422.

10. Anizen S, Milman G, Desrosiers N, Barnes AJ, Gorelick D, Huestis MA. Oral fluid cannabinoid concentrations following controlled smoked cannabis in chronic frequent and occasional smokers. *Anal Bioanal Chem.* 2013;405(26):8451−8461.

11. Newmeyer MN, Swortwood MJ, Andersson M, Abulseoud OA, Scheidweiler KB, Huestis MA. Cannabis edibles: blood and oral fluid cannabinoid pharmacokinetics and evaluation of oral fluid screening devices for predicting $\Delta^9$-tetrahydrocannabinol in blood and oral fluid following cannabis brownie administration. *Clin Chem.* 2017;63(3):647−662.

12. Vandrey R, Herrmann ES, Mitchell JM, et al. Pharmacokinetic profile of oral cannabis in humans: blood and oral fluid disposition and relation to pharmacodynamic outcomes. *J Anal Toxicol.* 2017;41(2):83−99.

13. Herrmann ES, Cone EJ, Mitchell JM, et al. Non-smokers exposure to secondhand smoke II: Effect of room ventilation on the physiological, subjective, and behavioral/cognitive effects. *Drug Alcohol Dep.* 2015;151:194−202.

14. Moore C, Coulter C, Uges D, et al. Cannabinoids in oral fluid following passive exposure to marijuana smoke. *For Sci Int.* 2011;212:227−230.

15. Niedbala RS, Kardos KW, Fritch DF, et al. Passive cannabis smoke exposure and oral fluid testing. II. Two studies of extreme cannabis smoke exposure in a motor vehicle. *J Anal Toxicol.* 2005;29(7):607−615.

16. White RM, Hart ED, Mitchell JM, SAMHSA Oral Fluid Pilot PT, 2011−2017.

17. Watanabe K, Yamaori S, Funahashi T, Kimura T, Yamamoto I. Cytochrome P450 enzymes involved in the metabolism of tetrahydrocannabinols and cannabinol by human hepatic microsomes. *Life Sci.* 2007;80:1415–1419.

18. Moore C, Rana S, Coulter C, Day D, Vincent M, Soares J. Detection of conjugated 11-nor-$\Delta^9$-tetrahydrocannabinol-9-carboxylic acid in oral fluid. *J Anal Toxicol.* 2007;31(4):187–194.

19. Mazur A, Lichti CF, Prather PL, et al. Characterization of human hepatic and extrahepatic UDP-glucuronosyl transferase enzymes involved in the metabolism of classic cannabinoids. *Drug Metab Disp.* 2009;37:1496.

20. Vondracek M, Xi Z, Larsson P, et al. Cytochrome P450 expression and related metabolism in human buccal mucosa. *Carcinogenesis.* 2001;22(3):481–488.

21. Yang S-P, Raner GM. Cytochrome *P*450 expression and activities in human tongue cells and their modulation by green tea extract. *Toxicol Appl Pharmacol.* 2005;202:140–150.

22. Lee D, Schwope DM, Milman G, Barnes AJ, Gorelick DA, Huestis MA. Cannabinoid disposition in oral fluid after controlled smoked cannabis. *Clin Chem.* 2012;58(4):748–756.

23. Baselt RC, ed. *Disposition of Toxic Drugs and Chemicals in Man.* 10th ed. Seal Beach, CA: Biomedical Publications; 2014, p. 511.

24. Ellefsen KN, Concheiro M, Pirard S, Gorelick DA, Huestis MA. Oral fluid cocaine and benzoylecgonine concentrations following controlled intravenous cocaine administration. *Forensic Sci Int.* 2016;260:95.

25. Baselt RC, ed. *Disposition of Toxic Drugs and Chemicals in Man.* 10th ed. Seal Beach, CA: Biomedical Publications; 2014, p. 122.

26. Baselt RC, ed. *Disposition of Toxic Drugs and Chemicals in Man.* 10th ed. Seal Beach, CA: Biomedical Publications; 2014, p. 1263.

27. Baselt RC, ed. *Disposition of Toxic Drugs and Chemicals in Man.* 10th ed. Seal Beach, CA: Biomedical Publications; 2014, p. 1286.

28. Baselt RC, ed. *Disposition of Toxic Drugs and Chemicals in Man.* 10th ed. Seal Beach, CA: Biomedical Publications; 2014, p. 1315.

29. Baselt RC, ed. *Disposition of Toxic Drugs and Chemicals in Man.* 10th ed. Seal Beach, CA: Biomedical Publications; 2014, p. 1318.

30. Gjerde H, Mordal J, Christopherson AS, Bramness JG, Mørland. Comparison of drug concentrations in blood and oral fluid collected with the Intercept® sampling device. *J Anal Toxicol.* 2010;34(4):204–209.

31. Schepers RJF, Oyler JM, Joseph Jr, et al. Methamphetamine and amphetamine pharmacokinetics in oral fluid and plasma after controlled oral methamphetamine administration to human volunteers. *Clin Chem.* 2003;49(1):121.

32. Huestis MA, Cone EJ. Methamphetamine disposition in oral fluid, plasma, and urine. *Ann N Y Acad Sci.* 2007;1098:104.

33. Barnes AJ, Scheidweiler KB, Kolbrich-Spargo EA, Gorlick DA, Goodwin RS. Huestis, MDMA and metabolite disposition in expectorated oral fluid after controlled oral MDMA administration. *Ther Drug Monit.* 2011;33(5):602–608.

34. Baselt RC, ed. *Disposition of Toxic Drugs and Chemicals in Man.* 10th ed. Seal Beach, CA: Biomedical Publications; 2014, p. 516.

35. Baselt RC, ed. *Disposition of Toxic Drugs and Chemicals in Man.* 10th ed. Seal Beach, CA: Biomedical Publications; 2014, p. 1399.

36. Baselt RC, ed. *Disposition of Toxic Drugs and Chemicals in Man.* 10th ed. Seal Beach, CA: Biomedical Publications; 2014, p. 1010.

37. Baselt RC, ed. *Disposition of Toxic Drugs and Chemicals in Man.* 10th ed. Seal Beach, CA: Biomedical Publications; 2014, p. 1017.

38. Baselt RC, ed. *Disposition of Toxic Drugs and Chemicals in Man.* 10th ed. Seal Beach, CA: Biomedical Publications; 2014, p. 1528.

39. Baselt RC, ed. *Disposition of Toxic Drugs and Chemicals in Man.* 10th ed. Seal Beach, CA: Biomedical Publications; 2014, p. 1534.

40. Baselt RC, ed. *Disposition of Toxic Drugs and Chemicals in Man.* 10th ed. Seal Beach, CA: Biomedical Publications; 2014, p. 992.

41. Kim I, Barnes AJ, Oyler JM, et al. Plasma and oral fluid pharmacokinetics and pharmacodynamics after oral codeine administration. *Clin Chem.* 2002;48(9):1486−1496.

42. Cone EJ, DePriest AZ, Heltsley R, et al. Prescription opioids. IV. Disposition of hydrocodone on oral fluid and blood following single dose administration. *J Anal Toxicol.* 2015;39 (7):510−518.

43. Cone EJ, DePriest AZ, Heltsley R, et al. Prescription opioids. III. Disposition of oxycodone on oral fluid and blood following single dose administration. *J Anal Toxicol.* 2015;39 (3):192−202.

44. Baselt RC, ed. *Disposition of Toxic Drugs and Chemicals in Man.* 10th ed. Seal Beach, CA: Biomedical Publications; 2014, p. 74.

45. Baselt RC, ed. *Disposition of Toxic Drugs and Chemicals in Man.* 10th ed. Seal Beach, CA: Biomedical Publications; 2014, p. 618.

46. Baselt RC, ed. *Disposition of Toxic Drugs and Chemicals in Man.* 10th ed. Seal Beach, CA: Biomedical Publications; 2014, p. 889.

47. Baselt RC, ed. *Disposition of Toxic Drugs and Chemicals in Man.* 10th ed. Seal Beach, CA: Biomedical Publications; 2014, p. 869.

48. Baselt RC, ed. *Disposition of Toxic Drugs and Chemicals in Man.* 10th ed. Seal Beach, CA: Biomedical Publications; 2014, p. 2037.

49. Baselt RC, ed. *Disposition of Toxic Drugs and Chemicals in Man.* 10th ed. Seal Beach, CA: Biomedical Publications; 2014, p. 846.

50. Bista SR, Lobb M, Haywood A, Hardy J, Tapuni A, Norris R. Development, validation and application of an HPLC-MS/MS method for the determination of fentanyl and nor-fentanyl in human plasma and saliva. *J Chromatogr B.* 2014;960:27−33.

51. Freye E. A new transmucosal drug delivery system for patients with breakthrough cancer pain: the fentanyl effervescent tablet. *J Pain Res.* 2008;2:13−20.

52. Daniulaityte R, Jehascik MP, Strayer KE, et al. Overdose deaths related to fentanyl and its analogues—Ohio, January-February 2017. *Morbid Mortal Weekly Rep.* 2017;66 (34):904−908.

53. Helland A, Brede WR, Michelsen LS, et al. Two hospitalizations and one death after exposure to ortho-fluorofentanyl. *J Anal Toxicol.* 2017;41(8):708−709.

54. Hilderbrand R. The fentanyl family of opioids. Drug Test Matters; 2013.

55. Baselt RC, ed. *Disposition of Toxic Drugs and Chemicals in Man.* 10th ed. Seal Beach, CA: Biomedical Publications; 2014, p. 61.

56. Baselt RC, ed. *Disposition of Toxic Drugs and Chemicals in Man.* 10th ed. Seal Beach, CA: Biomedical Publications; 2014, p. 1776.

57. Baselt RC, ed. *Disposition of Toxic Drugs and Chemicals in Man.* 10th ed. Seal Beach, CA: Biomedical Publications; 2014, p. 1882.

58. Heiskanen T, Langel K, Gunnar T, Lillsunde P, Kalso EA. Opioid concentrations in oral fluid and plasma in cancer patients with pain. *J Pain Symptom Manage*. 2015;50(4):524–532.

59. Bista SR, Haywood A, Norris R, et al. Saliva versus plasma for pharmacokinetic and pharmacodynamic studies of fentanyl in patients with cancer. *Clin Ther*. 2015;37(11):2468–2475.

60. Ruan X. Letter to the Editor concerning Bista SR et al. *Clin Ther*. 2015;37(11):2468. Clin Ther. 2015;37(12):28801.

61. Heltsley R, DePriest A, Black DL, et al. Oral fluid drug testing of chronic pain patients. I. Positive prevalence rates of licit and illicit drugs. *J Anal Toxicol*. 2011;35:529–540.

62. Davies A, Mundin G, Vriens J, Webber K, Buchanan A, Waghorn M. The influence of low salivary flow rates on the absorption of a sublingual fentanyl citrate formulation for breakthrough cancer pain. *J Pain Symptom Manage*. 2016;51(3):538–545.

63. Miller EI, Norris H-RK, Rollins DE, et al. Identification and quantification of nicotine biomarkers in human oral fluid from individuals receiving low-dose transdermal nicotine: a preliminary study. *J Anal Toxicol*. 2010;34(7):357. 2010.

64. Evans-Brown M, Gallegos A, Francis W, et al. New psychoactive substances in Europe. An update from the EU early warning system. <www.emcdda.europa.eu/system/files/publications/65/TD0415135ENN.pdf>; March 2015 Accessed November 11, 2017.

65. Williams M, Martin J, Galettis P. A validated method for the detection of 32 bath salts in oral fluid. *J Anal Toxicol*. 2017;41(8):659–669.

66. Coulter C, Garnier M, Moore C. Synthetic cannabinoids in oral fluid. *Anal Toxicol*. 2011;35(7):424–430.

67. Kneisel S, Speck M, Moosmann B, Corneille T, Butlin NG, Auwärter V. LC/ESI-MS/MS method for quantification of 28 synthetic cannabinoids in neat oral fluid and its application to preliminary studies on their detection windows. *Anal Bioanal Chem*. 2013;405(14):4691–4706.

68. Kneisel S, Auwärter V, Kempf J. Analysis of 30 synthetic cannabinoids in oral fluid using liquid chromatography-electrospray ionization tandem mass spectrometry. *Drug Test Anal*. 2012;5(8):657–659.

69. Personal communication, MedTox-LabCorp; 2016.

CHAPTER *3*

# Collection of Oral Fluid

## INTRODUCTION

One of the neglected variables in oral fluid testing for drugs and meta-bolites is the manner of collection.[1] The sample can be collected as an undiluted fluid via passive drool, expectoration (e.g., into a plastic or glass tube), or using a collection pad, which is predominantly cotton or synthetic fibers. Manual saliva collection using any type of pad causes stimulation of saliva, which may in turn affect both the pH of the fluid and the drug concentration.[2] Some pads are also treated with additives (e.g., citric acid) to stimulate saliva production. The pad then may or may not be placed into transportation buffer containing preser-vatives (surfactants, antibacterial components). Alternatively, neat oral fluid may be collected via a rinsing solution (e.g., Saliva Collection System; SCS). In 2008, Drummer published a review of collection techniques and applications of drug testing which noted "it is essential that devices used to collect oral fluid are checked to ensure reasonable stability and recovery of absorbed drug."[3] Almost 10 years later those scientific principles are still valid for evaluation of oral fluid collection devices.

The proposed Substance Abuse and Mental Health Services Administration (SAMHSA) guidelines for workplace drug testing suggest that a minimum of 1 mL ± 10% of neat oral fluid be collected regardless of device.[4] Both SAMHSA guidelines and the European Guidelines for Workplace Drug Testing in Oral Fluid recommend that either adequate volume for a split sample or consecutive samples be collected.[5]

Table 3.1 lists the main devices now commercially available to col-lect oral fluid that can be analyzed for drugs and metabolites. Over the last few years, many of these have been for sale, but not all are still in production. The list is not exhaustive but includes those most com-monly employed.

Detection of Drugs and Their Metabolites in Oral Fluid. DOI: https://doi.org/10.1016/B978-0-12-814595-1.00003-9

## Table 3.1 Oral Fluid Collection Devices

| Name | Mode of Collection | Volume Adequacy Indicator | Volume Collected (μL) | Manufacturer and Location |
|---|---|---|---|---|
| SalivaBio Passive Drool (with Saliva Collection Aid) | Neat | Yes | To mark | Salimetrics, CA |
| Salicule | Neat | Yes | To mark | Acro Biotech, CA |
| RapidEASE | Neat | Yes | To mark (2 mL) | Biophor Diagnostics Inc., CA |
| Saliva Split Collector | Neat | Yes | To mark | Sciteck, NC |
| UltraSal-2 | Neat | Yes | To mark | Neogen, KY |
| Saliva Collection System | Rinsing solution | No | Determine in laboratory | Greiner Bio-One, Austria |
| OnTrak Oratube | Pad, no buffer | No | Unknown at collection | Agilent Technologies, CA |
| Salivette | Pad, no buffer | No | 1.1 ± 0.3 mL | Sarstedt, Germany |
| Quantisal | Pad with buffer | Yes | 1000 ± 10% (Fig. 3.1) | Immunalysis, CA |
| Saliva Sampler | Pad with buffer | Yes | 1000 ± 10% | StatSure Diagnostic Systems, MA |
| Intercept | Pad with buffer | No | 400 (200–1800) | Orasure Technologies, PA |
| Intercept i2 and i2he | Pad with buffer | Yes | 1000 (Fig. 3.2) | Orasure Technologies, PA |
| Oral-Eze | Pad with buffer | Yes | 1000 | Thermo Fisher, CA |
| Aware Messenger | Pad with buffer | No | At least 1 mL | Calypte, OR |
| Certus | Pad with buffer | Yes | 650 | Alere Toxicology, UK |
| NeoSal | Pad with buffer | Yes | 700 (Fig. 3.3) | Neogen, KY |

*Figure 3.1 Quantisal device.*

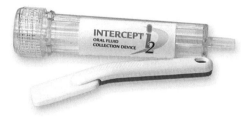

*Figure 3.2 Intercept.*

*Figure 3.3 NeoSal.*

## NEAT ORAL FLUID

Poor oral hygiene, smoking, and drug use can affect the production of saliva, and the time needed to collect adequate volume will vary between subjects. Saliva flow rates are usually lower when expectoration is used as the collection method; however, neat devices should still require that a minimum volume be collected to provide adequate fluid for analysis.

Passively collected drool is not widely used in drug testing, even though the expense of a collection device is eliminated. Expectorated samples have the advantage of being undiluted but may froth, making it difficult to see when an adequate volume has been collected. However, adequate data can be generated by this method. In one published study,[6] oral fluid was collected via nonstimulated expectoration into 50 mL polypropylene tubes, centrifuged at 1800 × $g$, and stored frozen ($-20°C$) until analysis for 3,4-methylenedioxymethamphetamine (MDMA, "ecstasy") and metabolites. In another study,[7] participants were asked to expectorate into polypropylene tubes. The sample had to be centrifuged and stored at $-20°C$ in Nunc cryotubes until analysis. A comparison of tetrahydrocannabinol (THC) stability between expectorated oral fluid and samples collected with a pad/buffer device (Quantisal) showed that cannabinoids in expectorated oral fluid were less stable than in Quantisal samples when refrigerated or frozen. The authors noted that "cannabinoid stability varied by analyte, collection method, and storage duration and temperature, and across participants. Collection with a device containing an elution/stabilization buffer, sample storage at 4°C, and analysis within 4 weeks was the preferred methodology in order to maximize result accuracy."[8]

## Advantages
- Inexpensive.
- Easy collection.
- No correction for dilution factors is required in the analysis.

## Disadvantages
- Collection is not hygienic.
- Collection is slow.
- Sample is viscous.
- Samples may be contaminated with food or other substances.
- Drug stability during transportation and storage can be problematic. THC in particular is not stable in neat oral fluid when stored in plastic bottles; silanized glass can be recommended for longer term storage. Neat collection protocols generally require immediate freezing of the specimen.
- Cellular debris and mucus are observed, so centrifugation is necessary prior to extraction and analysis. Multiple freeze–thaw cycles serve to break down mucopolysaccharides in the saliva, which reduces viscosity and improves pipetting accuracy in analysis.

Some neat oral fluid collectors provide only one specimen; others have a mechanism to divide the sample into two tubes in case a second or split sample is required. An uncentrifuged sample of neat oral fluid that has been allowed to sit undisturbed results in a sediment at the bottom of the tube (Fig. 6.2 in Chapter 6).

## SALIVA COLLECTION SYSTEM

The SCS is unusual, somewhat of a hybrid device between pure neat oral fluid, which suffers from drug stability issues, and pad/buffer devices, which involve measurement of drug recovery from the pads. SCS consists of a rinsing solution, an oral fluid extraction solution tube, an oral fluid collection beaker, and transfer tubes containing preservatives. The subject rinses the mouth for approximately 2 min with extraction solution, which contains a yellow dye (tartrazine). After rinsing, the solution and associated oral fluid are expectorated into the beaker. The sample is then transferred into one or two vials for laboratory testing. The tartrazine serves as an internal standard, so that the amount of oral fluid collected can be determined in the laboratory by spectrophotometry; this has been reported to be between 4.9 and 10.5 mL, ensuring an adequate sample for analysis.[9] The main disadvantage would appear to be possible swallowing of the rinsing solution by the subject, so that an inadequate volume of saliva is collected.

## PAD, NO BUFFER

Some collection pads are supplied with no buffer. The oral fluid is removed by compression of the pad (e.g., Oratube) or centrifugation (e.g., Salivette). The Salivette consists of a cotton roll which is inserted into the mouth for a period of time, then placed in a plastic tube, and sent to the test facility. The device can be used with a neutral cotton pad or a citric acid–treated cotton pad to stimulate saliva production.

There are several published studies in which drugs were measured in oral fluid following collection with the Salivette. Crouch et al. reported that codeine concentrations in expectorated oral fluid were consistently higher than those collected with a Salivette,[2] while a different group reported no difference in cocaine and metabolite concentrations between expectorated, stimulated pad, and neutral Salivette collection.[10]

### Advantages
- Inexpensive.
- Easy collection.
- No correction for dilution factors is required in the analysis.

## Disadvantages

- Volume collected is not known until after compression or centrifugation.
- Drugs in general and THC in particular adhere strongly to untreated collection pads and must be removed in a laboratory with organic solvents for improved drug recovery.
- THC detection rates in samples collected with the Salivette have been reported lower than in expectorated samples.[11]

## PAD AND BUFFER

The vast majority of oral fluid collections in workplace drug testing, pain management programs, and criminal justice settings use a pad and buffer device. The buffer usually contains detergents, surfactants, and stabilizers. From a hygienic point of view, systems using a pad are preferred to expectoration, and pad collection is faster.

The effect of the collection device on drug concentration has been assessed for several drugs, including codeine, amphetamine, MDMA, cocaine, THC, morphine, diazepam, and alprazolam,[2,12,13] and more recently antipsychotic drugs.[14] The researchers concluded that the method of collection affects the quantitative result, and, as expected, there is less variation intra-device than inter-device. Not all devices were used as directed by the manufacturers, especially in terms of pad residence time in the buffer prior to extraction.

One of the main issues with pad collection is the efficiency of drug release into the buffer. Recovery of drugs is affected by the volume and composition of the collection buffer, the time that the pad is allowed to remain in the solution, and transportation conditions. Kauert et al. (2006) reported that THC adsorbed strongly onto the Intercept pad. Drug recovery using the standard elution procedure was only $37.8\% \pm 9.4\%$ at a concentration of 10 ng/mL and $55.6\% \pm 1.0\%$ at 100 ng/mL ($n = 5$ each). The researchers had to add methanol to the pad to release 25% more THC.[15]

Wille et al. evaluated THC recovery from the StatSure, Quantisal, and Certus. THC was totally recovered from the StatSure pad after storage for 24 h at room temperature or 7 days at 4°C. The Quantisal showed a loss of 15%−25% over the same periods, and the Certus result was not reproducible.[16] Anizan et al. compared cannabinoid stability in the StatSure and Oral-Eze; their data indicated that the stabilizing transportation buffers in both devices provided good stability for most of the drugs.[17] Ventura et al. also studied the stability of drugs in transportation because of concerns with proficiency programs in which laboratories may participate.[18] A more in-depth discussion of variables within proficiency programs is given in Chapter 8.

Cohier et al. compared the Quantisal and Certus for opiates, cannabinoids, amphetamines, cocaine and its metabolites, methadone, and buprenorphine for recovery efficiency and drug stability. All drugs were recovered in excess of 90% for the Quantisal; refrigerated storage was preferred to frozen storage for most of them.[19] Quintela et al. determined the recovery of amphetamine, methamphetamine, oxazepam, codeine, morphine, cocaine, benzoylecgonine, THC, and methadone at three different concentrations using the Quantisal; recovery of all drugs was $>80\%$.[20] In a study comparing the concentration of zopiclone detected in oral fluid collected simultaneously using two different devices, Intercept and StatSure, the correlation between the two results was fairly poor, $r^2 = 0.35$. The authors concluded that "results indicate that the type of sampling device may significantly affect the analytical result for zopiclone in sampled oral fluid."[21]

## Advantages
- Fast and hygienic collection.
- Food or other contaminants are filtered out.
- Most devices have a volume adequacy indication.
- Buffers help to lower sample viscosity and support drug stability during transportation and storage.

## Disadvantages
- Dilution factors are necessary for quantitative results.
- Not all devices have a volume adequacy indication.
- Drug recovery from the collection pad may not be consistent or comparable across devices.
- Buffers may contribute to matrix effects and/or ion suppression in liquid chromatography-mass spectrometry laboratory analyses.

Stability of drugs during laboratory extraction and for long-term storage has been studied. While most drugs are fairly stable under standard conditions, fluorescent lighting has been shown to cause

THC losses of over 50%. In the dark, the loss of THC at room temperature was approximately 20% over 14 days. Furthermore, when plastic separators used to compress the collection pad to remove oral fluid were stored in place, THC concentration was reduced by almost 30% after 3 days and 60% after 14 days.[22]

## SUMMARY

Before the selection of a collection device for a specific purpose or project, the reader should be aware that not all devices are created equally. There are important differences that will affect the analytical result and therefore the reliability and accuracy of the data generated. As well as following the manufacturer's instructions (volume adequacy activation, pad residence time, transportation method, etc.), other post-collection and laboratory handling precautions should be noted.

## REFERENCES

1. Crouch DJ. Oral fluid collection: the neglected variable in oral fluid testing. *Forens Sci Int*. 2005;150(2−3):165−173.

2. O'Neal CL, Crouch DJ, Rollins DE, Fatah AA. The effects of collection methods on oral fluid codeine concentrations. *J Anal Toxicol*. 2000;24(7):536−542.

3. Drummer OH. Introduction and review of collection techniques and applications of drug testing of oral fluid. *Ther Drug Monit*. 2008;30(2):203−206.

4. SAMHSA, May 15, 2015 Federal Register, 80 FR 28053, Oral Fluid Mandatory Guidelines; Washington, DC, Government Printing Office.

5. Brcak M, Beck O, Bosch T, et al. European guidelines for workplace drug testing in oral fluid. *Drug Test Anal*. 2017:1−14. Available from: https://doi.org/10.1002/dta.2229.

6. Barnes AJ, Scheidweiler KB, Kolbrich-Spargo EA, Gorelick DA, Goodwin RS, Huestis MA. MDMA and metabolite disposition in expectorated oral fluid after controlled oral MDMA administration. *Ther Drug Monit*. 2011;33(5):602−608. Available from: https://doi.org/10.1097/FTD.0b013e3182281975.

7. Milman G, Barnes AJ, Schwope DM, et al. Cannabinoids and metabolites in expectorated oral fluid after 8 days of controlled around-the-clock oral THC administration. *Anal Bioanal Chem*. 2011;401(2):599−607. Available from: https://doi.org/10.1007/s00216-011-5066-4.

8. Lee D, Milman G, Schwope DM, Barnes AJ, Gorelick DA, Huestis MA. Cannabinoid stability in authentic oral fluid after controlled cannabis smoking. *Clin Chem*. 2012;58(7):1101−1109. Available from: https://doi.org/10.1373/clinchem.2012.184929.

9. Raggam RB, Santner BI, Kollroser M, et al. Evaluation of a novel standardized system for collection and quantification of oral fluid. *Clin Chem Lab Med*. 2008;46(2):2287−2291.

10. Scheidweiler KB, Spargo EA, Kelly TL, Cone EJ, Barnes AJ, Huestis MA. Pharmacokinetics of cocaine and metabolites in human oral fluid and correlation with plasma concentrations after controlled administration. *Ther Drug Monit*. 2010;32(5):628−637. Available from: https://doi.org/10.1097/FTD.0b013e3181f2b729.

11. Teixeira H, Proença P, Verstraete A, Corte-Real F, Vieira DN. Analysis of delta-9-tetrahydrocannabinol in oral fluid samples using solid-phase extraction and high-performance liquid chromatography-electrospray ionization mass spectrometry. *Forens Sci Int.* 2005;150(2−3):205−211.

12. Coucke LD, De Smet L, Verstraete AG. Influence of sampling procedure on codeine concentrations in oral fluid. *J Anal Toxicol.* 2016;40(2):148−152.

13. Langel K, Engblom C, Pehrsson A, Gunnar T, Ariniemi K, Lillsunde P. Drug testing in oral fluid-evaluation of sample collection devices. *J Anal Toxicol.* 2008;32(6):393−401.

14. Fisher DS, Beyer C, van Schalkwyk G, Seedat S, Flanagan RJ. Measurement of clozapine, norclozapine, and amisulpride in plasma and in oral fluid obtained using 2 different sampling systems. *Ther Drug Monit.* 2017;39(2):109−117.

15. Kauert GF, Iwersen-Bergmann S, Toennes SW. Assay of Delta9-tetrahydrocannabinol (THC) in oral fluid-evaluation of the OraSure oral specimen collection device. *J Anal Toxicol.* 2006;30(4):274−277.

16. Wille SM, Di Fazio V, Ramirez-Fernandez Mdel M, Kummer N, Samyn N. Driving under the influence of cannabis: pitfalls, validation, and quality control of a UPLC-MS/MS method for the quantification of tetrahydrocannabinol in oral fluid collected with StatSure, Quantisal, or Certus collector. *Ther Drug Monit.* 2013;35(1):101−111.

17. Anizan S, Bergamaschi MM, Barnes AJ, et al. Impact of oral fluid collection device on cannabinoid stability following smoked cannabis. *Drug Test Anal.* 2015;7(2):114−120.

18. Ventura M, Pichini S, Ventura R, et al. Stability of drugs of abuse in oral fluid collection devices with purpose of external quality assessment schemes. *Ther Drug Monit.* 2009;31(2):277−280.

19. Cohier C, Mégarbane B, Roussel O. Illicit drugs in oral fluid: evaluation of two collection devices. *J Anal Toxicol.* 2017;41(1):71−76.

20. Quintela O, Crouch DJ, Andrenyak DM. Recovery of drugs of abuse from the Immunalysis Quantisal oral fluid collection device. *J Anal Toxicol.* 2006;30(8):614−616.

21. Gjerde H, Øiestad EL, Oiestad AM, et al. Comparison of zopiclone concentrations in oral fluid sampled with Intercept oral specimen collection device and StatSure Saliva Sampler and concentrations in blood. *J Anal Toxicol.* 2010;34(9):590−593.

22. Moore C, Vincent M, Rana S, Coulter C, Agrawal A, Soares J. Stability of Delta(9)-tetrahydrocannabinol (THC) in oral fluid using the Quantisal collection device. *Forens Sci Int.* 2006;164(2−3):126−130.

# CHAPTER 4

# Point of Collection Testing

## INTRODUCTION

Devices for testing drugs in oral fluid at the point of collection or point of care (POC) have been available for many years. POC testing falls into two classifications: with small handheld portable devices and with larger bench-top instruments operated in a physician's office or similar setting. Newer developments are devices for the analysis of oral fluid from drivers at the roadside, and programmable Bio-Nano-Chip systems for detecting drugs in oral fluid with an emphasis on extending drug test profiles and improving sensitivity.[1,2] While workplace testing has historically been the site for most POC tests, wider application has taken place in pain management settings, rehabilitation clinics, and physicians' offices, where test results can be produced during a patient visit. With improvements in sensitivity, technology, and instrumentation, POC tests that use oral fluid as the matrix have become commercially available, and their utility in many areas of drug analysis is increasing. Table 4.1 compares POC tests to laboratory-based tests on key parameters.

POC tests are generally based on lateral flow immunoassay technology. In visually read devices, the absence of a line usually indicates a positive result (competitive drug binding) and the presence of a line indicates a negative result. The devices are prone to the same advantages and drawbacks associated with cross-reactivity and antibody selection as other immunoassays. Advantages include convenient sample collection, ease of use, rapid results, straightforward interpretation, and relatively low costs associated with implementation of a drug testing program. The limitations vary with device, as not all commercial products behave with the same degree of accuracy and reliability. Test strips are often made by the same manufacturer and then implanted into different POC cassettes and test devices. Many of the oral fluid rapid tests are urine-based assays diluted to have the sensitivity in the range required for saliva testing. This is particularly observed in the devices that claim wide test panels (e.g., ketamine, propoxyphene,

Detection of Drugs and Their Metabolites in Oral Fluid. DOI: https://doi.org/10.1016/B978-0-12-814595-1.00004-0

**Table 4.1 Comparison of POC Tests to Laboratory-Based Tests**

| Parameter | POC Handheld Devices | POC Bench-Top Instruments | Laboratory Based |
|---|---|---|---|
| Collection | Rapid | Rapid | Rapid |
| Personnel | Nontechnical personnel | Trained personnel | Trained personnel |
| Cost | Inexpensive | More expensive | Expensive |
| Time for Results | Minutes | Minutes to hours | Days |
| Ease of Use | Simple; minimal training required | More complex equipment; moderate training required | Complex instrumentation; high level training required |
| Drug Test Panel | Limited to device design | Limited to immunoassay panels | Wide range |
| Quality Control | No permanent record | Included with batch results | Included with batch results |
| Result Generation | Visual or instrumented | Instrumented | Instrumented |
| Result Retention | Some allow electronic retention | Electronically retained | Electronically retained |
| Proficiency Programs | None | Available | Available |

oxycodone, barbiturates, and buprenorphine). In oral fluid, the predominant drugs are the parent compounds [e.g., tetrahydrocannabinol (THC) and cocaine] and not the main urinary metabolites (11-nor-9-carboxy-delta 9-tetrahydrocannabinol (THC-COOH) and benzoylecgonine (BZE), respectively). Using a modified urine assay for oral fluid analysis does not target the correct drug; as a result, true positives could be missed when these tests are implemented. Marketing claims can also be misleading. For example, several device manufacturers market their marijuana detection level as relatively low (12 ng/mL), but in fact the target analyte for the immunoassay is the metabolite THC-COOH, which is present in oral fluid at 1000 times lower concentration than THC.

## EVALUATION

There are many commercially available oral fluid POC tests; information on 15 devices is presented in Table 4.2. There are significant differences in drug test panels, cutoff concentrations, and result interpretation and retention. Evaluation over the years has generally concluded that performance is variable: for some drugs, the tests are

## Table 4.2 Cutoff Concentrations for Various POC Devices for Detection of Drug Classes[a]

| Device | Results | Number of Drug Classes | Cutoff Concentrations (ng/mL) | | | | | | | | | |
|---|---|---|---|---|---|---|---|---|---|---|---|---|
| | | | THC/THC-COOH | AMP/METH/MDMA | Morphine | Cocaine BZE | Benzodiazepines | Methadone/EDDP | PCP | Oxycodone | Barbiturates | Buprenorphine |
| DrugTest 5000 | Print/Electronic | 7 | 5 | 50/35/100 | 20 | 20 | diazepam 15 | 20 | | | | |
| DDS2 | Print/Electronic | 6 | 25 | 50/50 | 40 | *30* | temazepam 20 | | | | | |
| DrugWipe 5 DrugWipe 6 (ketamine) | Visual/Reader | 5 or 6 | 10 | 25/10/10 | 25 | 10 | diazepam 10 | | | | | |
| OrAlert | Visual | 7 | 100 | 50/50 | 40 | *20* | oxazepam 10 | | 10 | | | |
| Oratect | Visual | 6 | 40 | 50/50 | 40 | 20 | | | 10 | | | |
| Oral-AQ 6 and 7 | Visual | 6 or 7 | 25 | 50/50/35 | 25 | 20 | oxazepam 5 or 10 | | 10 | | | |
| Rapid STAT | Visual/Reader | 6 | 15 | 25/25 | 25 | *12* | oxazepam 25 | | | | | |
| iScreen | Visual | 6 | *12* | 50/50 | 40 | *20* | | | 10 | | | |
| OralView | Visual | 8 | *12* | 50/50 | 40 | 20 | 20 | 30 | 10 | | | |
| Oral fluid cassette | Visual | 11 | *12* | 50/50 | 40 | 20 | oxazepam 50 | 35 | 10 | | | |
| StatSwab | Visual | 11 | *12* | 50/50 | 40 | *20* | oxazepam 50 | | 10 | 50 | 300 | 10 |
| Oraline/SalivaConfirm (Also includes fentanyl (10)) | Visual | 12 | THC 75 *THC-COOH 12* | 50/50 | 40 | 20 | 20 | 30 | 10 | 50 | 300 | 10 |
| Rapid Detect/Saliva Scan/Oral Cube (Also includes: cotinine (50), ketamine (50), propoxyphene (50)) | Visual | 15 | THC 100 *THC-COOH 12* | 50/50 | 40 | 20 | oxazepam 10 | methadone 30 *EDDP* 20 | 10 | 20 | 50 | 5 |

*Cutoff concentrations in italics are for the metabolite.*

AMP, amphetamine; METH, methamphetamine; MDMA, 3,4-methylenedioxymethamphetamine ("ecstasy"); BZE, benzoylecgonine; EDDP, 2-ethylidene-1,5-dimethyl-3,3-diphenylpyrrolidine; PCP, phencyclidine.

[a]Some manufacturers offer different cartridges with single or multiple drug classes, depending on the market requirements. The table is not exhaustive but seeks to give an overview of the types of products which are commercially available.

specific and reliable, and for others, predominantly marijuana and benzodiazepines, improvements in sensitivity are necessary.[3–5] Determination of a cutoff concentration (the concentration below which the result must be reported as negative) is problematic because research on saliva drug concentrations, while expanding, is much more limited than that on urine or blood. Professional societies recommend drug test levels, but usually these are for workplace testing and are not really relevant to either clinical applications or driving under the influence of drugs (DUID) cases.

In 2003, Walsh et al. evaluated six POC devices,[6] then an additional four in 2007.[7] The devices were OralLab, RapiScan, DrugWipe, SalivaScreen, Oratect, Uplink, OralStat, SmartClip, Impact, and OraLine IV s.a.; many of these are no longer available. Overall, the results were acceptable for methamphetamine and opiates, with differences between devices for cocaine and amphetamines. In the later paper, they reported that there were still several false-negative results, and that the sensitivity and performance of commercial devices was improving but remained problematic for the reliable detection of cannabinoid use. A much more recent paper indicated that the DrugWipe 5A when used in Italian discos, pubs, and music bars showed good sensitivity, specificity, and accuracy for amphetamines (80%), average sensitivity for cocaine (67%) and opiates (50%), and sensitivity of only 29% and accuracy of 53% for cannabis.[8]

## SPECIFIC APPLICATIONS

### Driving Under the Influence of Drugs

The preferred specimen for analysis in drugged driving identification is blood because blood more closely reflects the drug circulation in the body than urine. The disadvantage of blood analysis is the time necessary to collect the sample, which requires a medical professional and, in the United States, often a warrant. THC, the psychoactive compound in marijuana, is dissipated rapidly throughout the body; therefore, it is often not detected because of the time lapse between law enforcement stop and blood collection. Oral fluid, which is a reflection of the free drug circulation in the body, can be collected and analyzed at the roadside with commercially available POC devices. Australia was the first country to implement roadside saliva testing, using a device for the identification of THC and methamphetamine (with cross-reactivity to MDMA); other countries (e.g., the United Kingdom, Spain, Germany, and Belgium) have now introduced roadside tests with different drug tests panels and various devices.

In 2010, researchers evaluated three roadside devices in Belgian drivers.[9] Oral fluid drug screening was followed by plasma collection and confirmation. The group reported on the outcome in 2015, stating that "by changing the drug screening procedure and lowering the cut-off values for confirmation in plasma, the new law in Belgium has resulted in a better approach toward driving under the influence of drugs."[10]

Three oral fluid POC devices have been extensively evaluated in roadside settings by law enforcement and researchers: the DDS2 (Alere), the DrugTest 5000 (Draeger), and the DrugWipe (Securetec) (Fig. 4.1). Recently published evaluations have shown good overall

*Figure 4.1 Three commercially available oral fluid POC devices: DDS2, DrugTest 5000, DrugWipe (top to bottom).*

performance in terms of sensitivity, specificity, and accuracy. To be accepted in routine use, they must demonstrate validity, reliability, and robustness. A key provision for law enforcement use is the ability to print and/or retain the result; the DDS2 and DrugTest 5000 have such capability, but visually read tests do not.

For the analysis of cocaine in oral fluid, Brazilian researchers concluded that with a 10 ng/mL cutoff, the DDS2 "achieved reliability parameters higher than 80%."[11] An evaluation of the DrugTest 5000 for cocaine in a controlled setting also concluded that "the Draeger cocaine test-strip with cocaine only confirmation offers a useful option for monitoring the acute intoxication phase of DUID; additionally the [benzoylecgonine] test-strip with cocaine and/or [benzoylecgonine] confirmation increases the length of detection of cocaine intake for workplace drug testing, drug court, parole, pain management, drug treatment programs and both the acute cocaine intoxication and cocaine crash/fatigue phase of DUID."[12]

In the United States, police departments and researchers in Wisconsin,[13] Oklahoma,[14] and Kansas[15] evaluated the DDS2 in the field and concluded that it is a useful tool to assist law enforcement in identifying drugged drivers. The researchers noted that the test panel (six drug classes) may require expansion, as novel psychoactive substances are being used routinely, and that the sensitivity for benzodiazepines should be improved.

In 2014, Logan et al. evaluated the DrugWipe and DrugTest 5000 in a roadside experiment and concluded that they performed comparably, although the Drug Test 5000 was significantly better at detecting cannabis.[16]

All of these evaluations indicate that sensitivity to benzodiazepines must be improved because of their impairing nature as well as poor incorporation into oral fluid due to strong binding to proteins and weak acidity. Pehrsson et al. (2008) evaluated two different DrugWipe devices specifically focused on benzodiazepines; they also concluded that improved sensitivity is required.[17]

## Clinical Applications

Urine is the preferred specimen for testing patients in rehabilitation and pain management programs when establishing prescription compliance or illicit drug use. Oral fluid has a distinct advantage because it is minimally invasive and collection is observed, which, unlike urine collection, makes adulteration difficult. Studies in which paired oral fluid and urine collected from the same patients were analyzed to determine whether oral fluid provided similar information to urine have shown oral fluid to be a viable matrix for laboratory-based testing but not for POC.[18-20]

As more physician office laboratories are opened, smaller bench-top analyzers can also be considered as POC tests. Analysis at the clinic reduces turnaround time and expense associated with larger hospital laboratory testing. Oral fluid immunoassays can be placed on such instruments and operated while a patient is still present in the clinic; bench-top designs also have a wider drug test menu available. As with all screening techniques, positive results should be confirmed with a separate analytical test based on a different scientific principle (usually gas or liquid chromatography coupled to mass spectrometry).

A recent review article on rapid assessment of drugs of abuse included a section on oral fluid POC tests.[21] Laboratory-based testing obviously provides more accurate results than POC, but the convenience and relatively low cost of a reliable preliminary result are attractive factors in drug testing. POC oral fluid tests continue to improve, and reliable systems are increasingly commercially available. Instruments with biochip technology and/or enhanced detection techniques have been added to the marketplace in the last few years and should see wider application.

## REFERENCES

1. Christodoulides N, De La Garza R, Simmons GW, et al. Next generation programmable Bio-Nano-Chip system for on-site detection in oral fluids. *J Drug Abuse* 2015;1(1):1−6.

2. Christodoulides N, De La Garza R, Simmons GW, et al. Application of programmable Bio-Nano-Chip system for the quantitative detection of drugs of abuse in oral fluids. *Drug Alc Depend.* 2015;153:306−313.

3. Crouch DJ, Walsh JM, Cangianelli L, Quintela O. Laboratory evaluation and field application of roadside oral fluid collectors and drug testing devices. *Ther Drug Monit.* 2008;30 (2):188−195.

4. Vanstechelman S, Isalberti C, Van der Linden T, Pil K, Legrand S, Verstraete AG. Analytical evaluation of four on-site oral fluid drug testing devices. *J Anal Toxicol.* 2012;36(2):136−140.

5. Musshoff F, Hokam EG, Bott U, Madea B. Performance evaluation of on-site oral fluid drug screening devices in normal police procedure in Germany. *Forens Sci Int.* 2014;238:120−124.

6. Walsh JM, Flegel R, Crouch DJ, Cangianelli L, Baudys J. An evaluation of rapid point-of-collection oral fluid drug testing devices. *J Anal Toxicol*. 2003;27(7):429−439.

7. Walsh JM, Crouch D, Danaceau JP, Cangianelli L, Liddicoat L, Adkins R. Evaluation of ten oral fluid point-of-collection drug-testing devices. *J Anal Toxicol*. 2007;31(1):44−54.

8. Gentili S, Solimini R, Tittarelli R, et al. A study on the reliability of an on-site oral fluid drug test in a recreational context. *J Anal Methods Chem*. 2016;1234581. Available from: https://doi.org/10.1155/2016/1234581.

9. Wille SM, Samyn N, Ramírez-Fernández Mdel M, De Boeck G. Evaluation of on-site oral fluid screening using DrugWipe-5(+), RapidSTAT and Drug Test 5000 for the detection of drugs of abuse in drivers. *Forensic Sci Int*. 2010;198(1-3):2−6. Available from: https://doi.org/10.1016/j.forsciint.2009.10.012.

10. Van der Linden T, Wille SM, Ramírez-Fernandez M, Verstraete AG, Samyn N. Roadside drug testing: comparison of two legal approaches in Belgium. *Forensic Sci Int*. 2015;249:148−155. Available from: https://doi.org/10.1016/j.forsciint.2015.01.034.

11. Scherer JN, Fiorentin TR, Sousa TRV, Limberger RP, Pechansky F. Oral fluid testing for cocaine: analytical evaluation of two point-of-collection drugs screening devices. *J Anal Toxicol*. 2017;41(5):392−398. Available from: https://doi.org/10.1093/jat/bkx018.

12. Ellefsen KN, Concheiro M, Pirard S, Gorelick DA, Huestis MA. Cocaine and benzoylecgonine oral fluid on-site screening and confirmation. *Drug Test Anal*. 2016;8(3−4):296−303. Available from: https://doi.org/10.1002/dta.1966.

13. Edwards L, Smith KL, Savage T. Drugged driving in Wisconsin: oral fluid versus blood. *J Anal Toxicol*. 2017;41(6):523−529. Available from: https://doi.org/10.1093/jat/bkx051.

14. Veitenheimer AM, Wagner JR. Evaluation of oral fluid as a specimen for DUID. *J Anal Toxicol*. 2017;41(6):517−522. Available from: https://doi.org/10.1093/jat/bkx036.

15. Rohrig TP, Moore CM, Stephens K, et al. Road-side drug testing: an evaluation of the Alere DDS®2 mobile test system. *Drug Test Anal*. 2017:1−8. Available from: https://doi.org/10.1002/dta.2297 [Epub ahead of print].

16. Logan BK, Mohr AL, Talpins SK. Detection and prevalence of drug use in arrested drivers using the Dräger Drug Test 5000 and Affiniton DrugWipe oral fluid drug screening devices. *J Anal Toxicol*. 2014;38(7):444−450.

17. Pehrsson A, Gunnar T, Engblom C, Seppä H, Jama A, Lillsunde P. Roadside oral fluid testing: comparison of the results of DrugWipe 5 and DrugWipe benzodiazepines on-site tests with laboratory confirmation results of oral fluid and whole blood. *Forens Sci Int*. 2008;175 (2−3):140−148.

18. Cao JM, Ma JD, Morello CM, Atayee R, Best BM. Observations on hydrocodone and its metabolites in oral fluid specimens of the pain population: comparison with urine. *J Opioid Manage*. 2014;10(3):177−186. Available from: https://doi.org/10.5055/jom.2014.0206.

19. Conermann T, Gosalia A, Kabazie AJ, et al. Utility of oral fluid in compliance monitoring of opioid medications. *Pain Phys*. 2014;17:63−70.

20. Kunkel F, Fey E, Borg D, Stripp R, Getto C. Assessment of the use of oral fluid as a matrix for drug monitoring in patients undergoing treatment for opioid addiction. *J Opioid Manage*. 2015;11(5):435−442. Available from: https://doi.org/10.5055/jom.2015.0293.

21. Wiencek JR, Colby JM, Nichols JH. Rapid assessment of drugs of abuse. *Adv Clin Chem*. 2017;80:193−225. Available from: https://doi.org/10.1016/bs.acc.2016.11.003.

# CHAPTER 5

# Initial Testing

## INTRODUCTION

Whether for clinical or forensic drug and drug metabolite measurement, initial testing (also called screening) of oral fluid is used primarily to determine if individual drugs or their metabolites are present at a given cutoff. The initial testing process thus usually eliminates unnecessary confirmatory testing while simultaneously directing the laboratory to perform confirmatory testing when required. An exception is when a borderline negative is obtained and confirmatory testing is performed simply to determine the presence of even a small amount of drug or metabolite of interest to the requesting organization. Although confirmatory testing for a negative initial test is disallowed in most employment-related programs, it may be allowable and even encouraged in other programs (e.g., pain management, law enforcement, and probation).

In addition to directing the laboratory to the drugs and metabolites that require confirmatory testing, initial test results may indicate the amount by which an oral fluid sample should be diluted for valid confirmation results when confirmatory testing is to be performed initially ("range-finding").

Oral fluid initial testing is similar to urine initial testing but differs in three aspects: the need for greater sensitivity in oral fluid testing, the differences in the targeted analyte in some drug classes, and the substantial differences in the matrix.

Multiple methods for initial testing are available. Table 5.1 categorizes them.

Point of collection testing, also clinically referenced as point-of-care testing, is discussed in Chapter 4.

Detection of Drugs and Their Metabolites in Oral Fluid. DOI: https://doi.org/10.1016/B978-0-12-814595-1.00005-2

**Table 5.1 Oral Fluid Laboratory-Based Initial Test Methods**

| Method | Subcategories | Examples (Manufacturer and Test Name) |
|---|---|---|
| Immunoassay | ELISA[a,1] | Immunalysis, Neogen |
| | HEIA[b,1] | Immunalysis, OraSure |
| | CEDIA[c,1] | Thermo Fisher |
| Chromatography—mass spectrometry | Gas chromatography | See text |
| | Liquid chromatography | |

[a]Enzyme-linked immunosorbent assay
[b]Homogeneous enzyme immunoassay
[c]Cloned enzyme donor immunoassay

It is notable that most immunologic methods are paired with a specific oral fluid collector, and a collector from one manufacturer is generally not compatible with the method of another manufacturer. For example, performing an OraSure immunoassay on a Quantisal-collected oral fluid may not be successful unless the laboratory has properly re-validated their OraSure immunoassay for use with the Quantisal device.

## SENSITIVITY

Table 5.2 is an extract from the 2017 Federal Register, which defines the cutoffs required for urine drug and metabolite testing.

Similar requirements for oral fluid drug and metabolite testing from the 2017 Mandatory Guidelines for Federal Workplace Drug Testing Programs are presented in Table 5.3.[3]

From Tables 5.2 and 5.3, it is immediately obvious that the sensitivity required for oral fluid testing is several-fold greater than that needed for urine testing. The cutoff for parent THC in oral fluid (4 ng/mL) is 8% that of the metabolite THCA (also called THC-COOH) in urine (50 ng/mL). The cutoff for cocaine's metabolite benzoylecgonine (BZE) in oral fluid (15 ng/mL) is 10% that for urine (150 ng/mL). For opioids, the comparison is more spectacular. The codeine and morphine cutoff is 30 ng/mL in oral fluid, 2000 ng/mL in urine, or 1.5%. The cutoff for the more pharmacodynamically potent hydrocodone and hydromorphone is 30 ng/mL in oral fluid, 300 ng/mL in urine, or 10%. For the even more potent oxycodone and oxymorphone, the numbers are 30 ng/mL and 100 ng/mL, or 33%. For the major metabolite of heroin, 6-acetylmorphine, the cutoff differential is 30%. For

## Table 5.2 Extract from the Federal Register Citing Urine Cutoffs[2]

Section 3.4 What are the drug test cutoff concentrations for urine?

| Initial Test Analyte | Initial Test Cutoff [a] | Confirmatory Test Analyte | Confirmatory Test Cutoff Concentration |
|---|---|---|---|
| Marijuana metabolites (THCA)[b] | 50 ng/mL[c] | THCA | 15 ng/mL |
| Cocaine metabolite (Benzoylecgonine) | 150 ng/mL[c] | Benzoylecgonine | 100 ng/mL |
| Codeine/Morphine | 2000 ng/mL | Codeine | 2000 ng/mL |
| | | Morphine | 2000 ng/mL |
| Hydrocodone/ Hydromorphone | 300 ng/mL | Hydrocodone Hydromorphone | 100 ng/mL |
| | | | 100 ng/mL |
| Oxycodone/Oxymorphone | 100 ng/mL | Oxycodone Oxymorphone | 100 ng/mL |
| | | | 100 ng/mL |
| 6-Acetylmorphine | 10 ng/mL | 6-Acetylmorphine | 10 ng/mL |
| Phencyclidine | 25 ng/mL | Phencyclidine | 25 ng/mL |
| Amphetamine/ Methamphetamine | 500 ng/mL | Amphetamine | 250 ng/mL |
| | | Methamphetamine | 250 ng/mL |
| MDMA[d]/MDA[e] | 500 ng/mL | MDMA MDA | 250 ng/mL |
| | | | 250 ng/mL |

[a] *For grouped analytes (i.e., two or more analytes that are in the same drug class and have the same initial test cutoff).*
*Immunoassay*: The test must be calibrated with one analyte from the group identified as the target analyte. The cross-reactivity of the immunoassay to the other analyte(s) within the group must be 80 percent or greater; if not, separate immunoassays must be used for the analytes within the group.
*Alternate technology*: Either one analyte or all analytes from the group must be used for calibration, depending on the technology. At least one analyte within the group must have a concentration equal to or greater than the initial test cutoff or, alternatively, the sum of the analytes present (i.e., equal to or greater than the laboratory's validated limit of quantification) must be equal to or greater than the initial test cutoff.
[b] *An immunoassay must be calibrated with the target analyte, Δ-9-tetrahydrocannabinol-9-carboxylic acid (THCA).*
[c] Alternate technology (THCA and benzoylecgonine): *The confirmatory test cutoff must be used for an alternate technology initial test that is specific for the target analyte (i.e., 15 ng/mL for THCA, 100 ng/mL for benzoylecgonine).*
[d] *Methylenedioxymethamphetamine (MDMA).*
[e] *Methylenedioxyamphetamine (MDA).*

amphetamine and methamphetamine, the numbers are 25 ng/mL and 500 ng/mL, or 5%. The same general relationship applies to the psychoto-mimetic methylenedioxy compounds methylenedioxymethamphetamine (MDMA) and methylenedioxyamphetamine (MDA).

Homogeneous enzyme immunoassays (HEIAs) such as enzyme-multiplied immunoassay technique (EMIT) and cloned enzyme donor immunoassay (CEDIA) have adequate sensitivity for urine testing and in some cases for blood and blood products. The same

**Table 5.3 Extract from the Federal Register Citing Proposed Oral Fluid Cutoffs[3]**

Section 3.4 What are the drug test cutoff concentrations for undiluted (neat) oral fluid?

| Initial Test Analyte | Initial Test Cutoff (ng/mL) | Confirmatory Test Analyte | Confirmatory Test Cutoff Concentration (ng/mL) |
|---|---|---|---|
| Marijuana (THC)[a] | 4 | THC | 2 |
| Cocaine/ Benzoylecgonine[b] | 15 | Cocaine | 8 |
| | | Benzoylecgonine | 8 |
| Codeine/Morphine[b] | 30 | Codeine | 15 |
| | | Morphine | 15 |
| Hydrocodone/ Hydromorphone[b] | 30 | Hydrocodone | 15 |
| | | Hydromorphone | 15 |
| Oxycodone/ Oxymorphone[b] | 30 | Oxycodone | 15 |
| | | Oxymorphone | 15 |
| 6-Acetylmorphine | 3 | 6-Acetylmorphine | 2 |
| Phencyclidine | 3 | Phencyclidine | 2 |
| Amphetamine/ Methamphetamine[b] | 25 | Amphetamine | 15 |
| | | Methamphetamine | 15 |
| MDMA[b,c]/MDA[d] | 25 | [c]MDMA | 15 |
| | | [d]MDA | 15 |

[a]$\Delta$-9-Tetrahydrocannabinol (THC).
[b]Immunoassay: The test must be calibrated with one analyte from the group identified as the target analyte. The cross reactivity of the immunoassay to the other analyte(s) within the group must be 80 percent or greater; if not, separate immunoassays must be used for the analytes within the group.
Alternate technology: Either one analyte or all analytes from the group must be used for calibration, depending on the technology. At least one analyte within the group must have a concentration equal to or greater than the initial test cutoff or, alternatively, the sum of the analytes present (i.e., equal to or greater than the laboratory's validated limit of quantification) must be equal to or greater than the initial test cutoff.
[c]Methylenedioxymethamphetamine (MDMA).
[d]Methylenedioxyamphetamine (MDA).
Methylenedioxyethylamphetamine was included in the first draft but most likely will be excluded in the final draft. As of this writing, the final OF Mandatory Guidelines have not been published.

may apply to oral fluid testing. The necessary additional sensitivity can be attained by the use of enzyme-linked immunosorbent assay (ELISA). The enhanced sensitivity of ELISA may be adversely counterbalanced by its lower suitability in a high test volume laboratory, compared to a homogeneous immunoassay. Chromatography with mass spectrometry, especially liquid chromatography—tandem mass spectrometry (LC-MS/MS), offers even more sensitivity than ELISA. However, the equipment is extremely expensive and the bench techniques are more labor-intensive.

## CROSS-REACTIVITY

Regardless of the matrix, the individual tasked with the development of an initial immunoassay is challenged by the issue of cross-reactivity. A similar issue, the inclusion or recognition of drugs and their metabolites from a given class for a chromatographic—mass spectrometric procedure, is addressed under "Chromatography—Mass Spectrometry."

The issue of cross-reactivity is difficult to resolve to the satisfaction of all end users. For a specific quantitative oral fluid assay such as that for the hormone cortisol, very little if any cross-reactivity with other corticosteroids is probably the manufacturer's ultimate goal. Attaining the lowest cross-reactivity with other nonstructurally related endobiologics ("background noise") also is highly desirable, as it is for drug class screens. However, when an oral fluid sample is screened for a class of drugs and their metabolites, the allowable and required cross-reactivity may become less obvious. As an example, a screening immunoassay for opioids should not detect only morphine except under highly restricted circumstances. Other opioids such as codeine, hydrocodone, and hydromorphone need to have sufficient cross-reactivity even if morphine is the calibrator, so that other opioids will be detected in the screen and confirmed as specific drugs or metabolites in further testing.

The specificity of an immunoassay is almost entirely dependent on the specificity of the antibody or antibodies used. Although a complete review of the generation of antibodies and their use in immunoassay kits is beyond the scope of this book, a few basic comments are in order. Antibodies used in drug testing kits are almost always either polyclonal or monoclonal. Although a drug such as morphine will react with an antibody to morphine, morphine by itself is usually too small to be capable of acting as an immunogen, that is, generating an immune response which results in the production of an antibody. However, morphine can be chemically connected to a carrier protein to become an immunogenic conjugate. The injection of the morphine—protein conjugate into an immunocompetent animal such as a sheep will produce antibodies to morphine.

However, morphine has to be functionalized with reactive groups such as amine ($NH_2$) or carboxyl (COOH) before binding to a carrier protein. Morphine derivatized with $NH_2$ or COOH is referenced as a hapten. When the morphine (or other drug or metabolite) antibody is derived from a specific cell line (i.e., one specific presentation of the morphine in the complex), the antibody is monoclonal.[1]

The needs of all drug testing end-users may not be met with one type of initial test. Table 5.4 is a summary of possibly desirable cross-reactivities in oral fluid immunoassays taken from the authors' experience.

An example of the broad differences in cross-reactivity between parent THC and the metabolite THC-COOH (also called THCA) is presented in Table 5.5.

| Table 5.4 Potentially Useful Cross-Reactivities for Oral Fluid Immunoassays | | |
|---|---|---|
| **Drug or Class** | **Calibrator** | **Cross-Reactivity** |
| Alcohol | Ethanol | Not applicable. Enzymatic assay only. |
| Cannabinoids | $\Delta^9$-THC (THC) | Antibody needs to be very specific for THC. Broader cross-reactivity may be useful if the identification of other active cannabinoids such as 11-hydroxy and 8-β-hydroxytetrahydrocannabinol is beneficial to a given program. Cross-reactivity with 11-nor-$\Delta^9$-tetrahydrocannabinol-9-carboxylic acid (THC-COOH or THCA) probably is not useful due to the marked disparity between levels of the parent drug and metabolite in human oral fluid. |
| Cocaine | Cocaine or BZE | Good cross-reactivity between the parent drug and its major metabolite, BZE, is useful to determine cocaine use after the disappearance of the parent drug from oral fluid. Conversely, good cross-reactivity with both cocaine and BZE is essential during the period immediately after use when cocaine is present in oral fluid almost exclusively as the parent with little or no metabolite. Almost 25% of cocaine-positive oral fluid specimens have been shown to contain cocaine only (no BZE).[4] |
| Amphetamine | *d*-Amphetamine | Low cross-reactivity with over-the-counter amphetamine analogues such as norephedrine and norpseudoephedrine. Cross-reactivity with scheduled amphetamine analogues such as MDA and psychoactive, ring-substituted analogues is desirable, especially where their identification is allowed or required in a given program. |
| Methamphetamine | *d*-Methamphetamine | Low cross-reactivity with *l*-methamphetamine and over-the-counter methamphetamine analogues such as ephedrine and pseudoephedrine. Cross-reactivity with scheduled methamphetamine analogues such as MDMA and psychoactive, ring-substituted analogues is desirable, especially where their identification is allowed or required in a given program. |
| MDMA | MDMA | Good cross-reactivity with the metabolite and separately marketed drug MDA. |

*(Continued)*

## Table 5.4 (Continued)

| Drug or Class | Calibrator | Cross-Reactivity |
|---|---|---|
| Opiates | Morphine | Good cross-reactivity with commonly encountered non-morphine opiates such as codeine, hydrocodone, and hydromorphone. Good cross-reactivity with oxycodone, oxymorphone, and their metabolites probably is not possible at the time of this writing. Thus, oxymorphone and oxycodone will need to be considered as a separate set of analytes. It is unknown at the time of this writing whether conjugates such as codeine and morphine glucuronides will become important. |
| Oxycodone and Oxymorphone | Oxycodone and/or oxymorphone | Good cross-reactivity between oxycodone and oxymorphone, which, at the time of this writing, is almost 100% in some commercial assays. Cross-reactivity with the nor-metabolites may be important if the laboratory includes them in their confirmatory menu. It is unknown at the time of this writing whether conjugates such as oxymorphone glucuronide will become important. |
| 6-Acetylmorphine | 6-Acetylmorphine | Low- to non-existent cross-reactivity with codeine, morphine, hydrocodone, hydromorphone, oxymorphone, and oxycodone and their major metabolites is essential to reducing false positives. |
| Phencyclidine | Phencyclidine | Specificity for phencyclidine and cross-reactivity limited to its chemical analogues where such analogues can be analyzed by the laboratory's confirmatory procedures and reported. |
| Fentanyl | Fentanyl | Low background noise due to the low level of the analyte, its metabolites, and chemical analogues. Should demonstrate cross-reactivity with common chemical analogues. |
| Benzodiazepines | Nordiazepam, nitrazepam, i.a. | Where numerous parent drugs and their metabolites can be measured by the testing laboratory, broad cross-reactivity for benzodiazepines as a class with low cross-reactivity toward chemically similar drugs. Low background due to the low level of the analyte. At the time of this writing it is unknown whether cross-reactivity toward conjugates such as glucuronides will be required. |
| Nicotine/cotinine | Cotinine | Specific assay with low cross-reactivity to other drugs or metabolites of nicotine. |

## Table 5.5 Oral Fluid Cannabinoid Immunoassay Cross-Reactivities

| Manufacturer and Assay Name | Calibrator | THC-COOH Cross-Reactivity |
|---|---|---|
| Immunalysis, Enzyme Immunoassay[5] | THC, 8 ng/mL | 150% |
| Immunalysis, ELISA[6] | THC, 4 ng/mL | 25% |
| Neogen, THC Oral Fluid[7] | $\Delta^9$-THC | 138% (-)-THC-COOH |
| OraSure, Intercept i2he Cannabinoids Oral Fluid[8] | $l$-$\Delta^9$-THC, 3 ng/mL | Positive, 4 ng/mL |
| Thermo, CEDIA Cannabinoids OFT[9] | $l$-$\Delta^9$-THC, 3 ng/mL | Positive, 3 ng/mL |

To obtain the required sensitivity for oral fluid testing, some laboratories have resorted to the use of chromatography (liquid or gas) coupled with mass spectrometry, a combination usually reserved for confirmatory testing in employment-related laboratories. However, again, the tradeoff for enhanced sensitivity using a technique such as LC-MS/MS or gas chromatography—mass spectrometry (GC-MS) is that chromatography—mass spectrometry may not be as well suited for high volume test production as in immunoassay.

## CHROMATOGRAPHY—MASS SPECTROMETRY

Chromatographic—mass spectrometric methods for initial testing may be divided into two approximate categories: (1) those that test a broad range of usually chemically unrelated compounds and (2) those that target a specific class. Allen et al.[10] compared an LC-MS/MS procedure to ELISA using 72 samples of oral fluid from an addiction clinic. They concluded that LC-MS/MS is more flexible, specific, and sensitive than ELISA. However, a high-volume testing laboratory that converts from ELISA to LC-MS/MS might find that they have replaced one high test volume-incompatible procedure with another.

In 2003, Fucci et al.[11] demonstrated that drugs and metabolites that differ markedly in their physical properties (e.g., methadone, 2-ethyl-1,5-dimethyl-3,3-diphenylpyrollidine, amphetamines, MDAs, cocaine, cocaethylene, cannabidiol, THC, and cannabinol) can be analyzed simultaneously after solid-phase extraction from saliva by GC-MS. Contemporaneously, Dams et al.[12] used LC-atmospheric pressure chemical ionization—MS/MS to analyze an even larger group of drugs and their metabolites with only protein precipitation using acetonitrile. Groups of multiple drugs and drug metabolites from oral fluid have been analyzed successfully by subsequent researchers.[13–30] Since synthetic cannabinoids (e.g., JWH compounds) are difficult to detect in urine due to extensive metabolism, Kneisel and co-workers[31,32] used both neat oral fluid and oral fluid collected by the Dräger DrugWipe device to detect multiple synthetic cannabinoids. It was noted that a sufficient amount of the synthetic cannabinoid to demonstrate the recent use exists probably due to residual drug from smoking, a situation not unlike the analysis of parent THC in oral fluid. "Spice" and stimulants have been addressed in a separate study that involved ultra high-performance liquid chromatography and electrospray ionization tandem mass spectrometry.[33] Benzodiazepines as a group also can be screened to determine the exact compound and its metabolites.[34,35]

Similarly to immunoassay, chromatographic–mass spectrometric assays can experience interference from substances that are not the targeted analyte or its internal standard. If the laboratory has the ability to confirm a drug or its metabolite not included in the initial test calibrators and drugs not included in the initial test menu, a finding of a drug not normally anticipated can be of major importance to a report such as that for a DUID (driving under the influence of drugs) criminal case if validated and allowed by the customer.

## CONCLUSIONS

At the time of this writing, there are several initial test options that are immunoassay or chromatographic–mass spectrometric methods. The choice depends on the regulations under which a laboratory operates, the laboratory's workflow, and the needs of the end user.

## REFERENCES

1. Kricka LJ, Park JY. In: Rifai N, Horvath AR, Wittwer CT, eds. *Immunochemical Techniques in Tietz Textbook of Clinical Chemistry and Molecular Diagnostics.* 6th ed. Elsevier; 2018.

2. Mandatory Guidelines for Federal Workplace Drug Testing Programs, Substance Abuse and Mental Health Services, Department of Health and Human Services, 82 FR 7920-7970; 2017.

3. Mandatory Guidelines for Federal Workplace Drug Testing Programs, Substance Abuse and Mental Health Services, Department of Health and Human Services, 80 FR 28053-28101; 2015.

4. Flood JG, Khaliq T, Bishop KA, Griggs DA. The new substance abuse and mental health services administration oral fluid cutoffs for cocaine and heroin-related analytes applied to an addiction medicine setting: Important, unanticipated findings with LC-MS/MS. *Clin Chem.* 2016;62(5):773–780. Available from: https://doi.org/10.1373/clinchem.2015.251066.

5. THC Oral Fluid Enzyme Immunoassay, Revision E.1. Pomona, CA: Immunalysis Corporation; 2015.

6. Saliva/Oral Fluids Cannabinoids ELISA Kit, Revision I.O. Pomona, CA: Immunalysis Corporation; 2016.

7. THC Oral Fluid Kit, D120519. Lansing, MI: Neogen Corporation; 2015.

8. Intercept® i2he™ Cannabinoids Oral Fluid Assay, 10019861-0. Bethlehem, PA: OraSure Technologies, Inc.; 2014.

9. CEDIA® Cannabinoids OFT Assay, 10014911-2. Fremont, CA: Microgenics Corporation; 2014.

10. Allen KR, Azad R, Field HP, Blake DK. Replacement of immunoassay by LC tandem mass spectrometry for the routine measurement of drugs of abuse in oral fluid. *Ann Clin Biochem.* 2005;42:277–284.

11. Fucci N, de Giovanni N, Chiarotta M. Simultaneous detection of some drugs of abuse in saliva samples by SPME technique. *Forensic Sci Int.* 2003;134:40–45.

12. Dams R, Murphy CM, Choo RE, Lambert WE, de Leenheer AP, Huestis MA. LC-atmospheric pressure chemical ionization-MS/MS analysis of multiple illicit drugs, methadone, and their metabolites in oral fluid following protein precipitation. *Anal Chem.* 2003;75:798–804.

13. Huestis MA. A new ultraperformance-tandem mass spectrometry oral fluid assay for 29 illicit drugs and medications. *Clin Chem.* 2009;55(12):2079–2081.

14. Concheiro M, de Castro A, Quintela Ó, Cruz A, López-Rivadulla M. Determination of illicit and medicinal drugs and their metabolites in oral fluid and preserved oral fluid by liquid chromatography-tandem mass spectrometry. *Anal Bioanal Chem.* 2008;391:2329–2338.

15. Wang I-T, Feng Y-T, Chen C-Y. Determination of 17 illicit drugs in oral fluid using isotope dilution ultra-high performance liquid chromatography/tandem mass spectrometry with three atmospheric pressure ionizations. *J Chromatogr B.* 2010;878:3095–3105.

16. Song SM, Marriott P, Wynne P. Comprehensive two-dimensional gas chromatography-quadrupole mass spectrometric analysis of drugs. *J Chromatogr A.* 2004;1058:223–232.

17. Gunnar T, Arinemi K, Lillsunde P. Validated toxicological determination of 30 drugs of abuse as optimized derivatives in oral fluid by long column fast gas chromatography/electron impact mass spectrometry. *J Mass Spectrom.* 2005;40:739–753.

18. Langel K, Gunnar T, Ariniemi K, Rajamäki O, Lillsunde P. A validated method for the detection and quantitation of 50 drugs of abuse and medicinal drugs in oral fluid by gas chromatography-mass spectrometry. *J Chromatogr B.* 2011;879:859–870.

19. Badawi N, Simonsen KW, Steentoft A, Bernhoft IM, Linnet K. Simultaneous screening and quantification of 29 drugs of abuse in oral fluid by solid-phase extraction and ultraperformance LC-MS/MS. *Clin Chem.* 2009;55(11):2004–2018.

20. Strano-Rossi S, Colamonici C, Botrè F. Parallel analysis of stimulants in saliva and urine by gas chromatography/mass spectrometry: perspectives for "in competition" anti-doping analysis. *Anal Chim Acta.* 2008;606:217.

21. Di Corcia D, Lisi S, Gerace E, Salamone A, Vincenti M. Determination of pharmaceutical and illicit drugs in oral fluid by ultra-high performance liquid chromatography-tandem mass spectrometry. *J Chromatogr B.* Online. 2013.

22. Mortier KA, Maudens KE, Lambert WE, et al. Simultaneous, quantitative determination of opiates, amphetamines, cocaine and benzoylecgonine in oral fluid by liquid chromatography quadrupole-time-of-flight mass spectrometry. *J Chromatogr B.* 2002;779:321.

23. Wylie FM, Torrance H, Anderson RA, Oliver JS. Drugs in oral fluid. Part I. Validation of an analytical procedure for licit and illicit drugs in oral fluid. *Forensic Sci Int.* 2005;150:191.

24. Øiestad EL, Johansen U, Christopheren AS. Drug screening of preserved oral fluid by liquid chromatography-tandem mass spectrometry. *Clin Chem.* 2007;53(2):300.

25. Concheiro M, Gray TR, Shakleya DM, Huestis MA. High-throughput simultaneous analysis of buprenorphine, methadone, cocaine, opiates, nicotine, and metabolites in oral fluid by liquid chromatography tandem mass spectrometry. *Anal Bioanal Chem.* 2010;398:915.

26. Sergi M, Compagnone D, Curini R, et al. Micro-solid phase extraction coupled with high-performance liquid chromatography-tandem mass spectrometry for the determination of stimulants, hallucinogens, ketamine and phencyclidine in oral fluids. *Anal Chim Acta.* 2010;675:132.

27. Wood M, Laloup M, del Mar Ramirez Fernandez M, et al. Quantitative analysis of multiple illicit drugs in preserved oral fluid by solid-phase extraction and liquid chromatography-tandem mass spectrometry. *Forensic Sci Int.* 2005;150:227.

28. Fritch D, Blum K, Nonnemacher S, Haggerty BJ, Sullivan MP, Cone EJ. Identification and quantitation of amphetamines, cocaine, opiates, and phencyclidine in oral fluid by liquid chromatography mass spectrometry. *J Anal Toxicol.* 2009;33:569.

29. Liu H-C, Lee H-T, Hsu Y-C, et al. Direct injection LC-MS/MS analysis of opiates, methamphetamine, buprenorphine, methadone and their metabolites in oral fluid from substitution therapy patients. *J Anal Toxicol.* 2015;39:472.

30. Yonamine M, Tawil N, de Moraes Moreau R, Sliva OA. Solid-phase micro-extraction-gas chromatography-mass spectrometry and headspace-gas chromatography of tetrahydrocannabinol, amphetamine, methamphetamine, cocaine and ethanol in saliva samples. *J Chromatogr B.* 2003;789:73.

31. Kneisel S, Speck M, Moosmann B, Corneille TM, Butlin NG, Auwärter V. LC/ESI-MS-MS method for quantification of 28 synthetic cannabinoids in neat oral fluid and its application to preliminary studies on their detection windows. *Anal Bioanal Chem.* 2013.

32. Kneisel S, Auwärter V, Kempf J. Analysis of 30 synthetic cannabinoids in oral fluid using liquid chromatography-electrospray ionization tandem mass spectrometry. *Drug Test Anal.* 2012.

33. Strano-Rossi S, Anzillotti L, Castrinanó E, Romolo FS, Chiarotti M. Ultra high performance liquid chromatography-electrospray ionization-tandem mass spectrometry-screening method for direct analysis of designer drugs, "spice" and stimulants in oral fluid. *J Chromatogr A.* 2012;1258:37.

34. Moore C, Coulter C, Crompton K, Zumwalt M. Determination of benzodiazepines in oral fluid using LC-MS-MS. *J Anal Toxicol.* 2007;31:596.

35. Ngwa G, Fritch D, Blum K, Newland G. Simultaneous analysis of 14 benzodiazepines in oral fluid by solid phase extraction and LC-MS-MS. *J Anal Toxicol.* 2007;31:369.

# CHAPTER 6

# Specimen Validity Testing

## INTRODUCTION

Specimen validity testing (SVT) is the analysis of selected endobiological parameters of a submitted donor specimen to ensure that they are within the limits usually considered normal, thus ensuring that the specimen originated from the donor and is not diluted, adulterated, or substituted. SVT is well established for employment-related urine drug testing.[1] As stated in Chapter 3, the collection of oral fluid for drug and metabolite testing should be observed, so that there is little opportunity for a donor to suborn a test by diluting or adulterating the specimen or substituting a drug-free product. Following are some methods by which the donor might attempt to suborn the test:

- Coerce or compensate the collector, monetarily or otherwise, to be allowed to alter the specimen by dilution or addition of chemicals deleterious to the testing process.
- Coerce or compensate the collector to be allowed to substitute the submitted specimen with a drug-free specimen that appears to be oral fluid.
- Place in the oral cavity a substance or substances that will interfere with laboratory analysis.
- Place in the oral cavity a substance or substances that will dilute the specimen so that the drug or metabolites will be below the cutoff for a positive.

The test could also be invalidated if collection is performed incorrectly.

## ORAL FLUID SVT

Although the likelihood of successful tampering with an oral fluid specimen is much less than with a urine specimen, it is still sufficient to justify some degree of SVT.

Detection of Drugs and Their Metabolites in Oral Fluid. DOI: https://doi.org/10.1016/B978-0-12-814595-1.00006-4

Human immunoglobulin G (IgG), presumably derived primarily from crevicular fluid, has been used as an SVT analyte.[2] It was found to produce acceptable assessment of specimen validity in the Oral Fluid Pilot PT Program of the Substance Abuse and Mental Health Services Administration (SAMHSA), which is part of the US Department of Health and Human Services.[3] This program also found oral fluid albumin to be an acceptable analyte.[3] However, IgG was deemed ineffective in at least one publication.[4]

In an effort to identify an effective oral fluid SVT marker, Friesen et al.[5] collected neat oral fluid (Biophor RapidEASE silanized with Sylon-CT or Supelco 33065-U) and oral fluid in three pad-type collectors (first-generation Intercept, Oral-Eze, and second-generation Quantisal) from five donors. Fluid was collected in accordance with each manufacturer's instructions and by inoculation of each type of pad with the amount specified by its manufacturer, followed by insertion of the pad into the manufacturer's buffer/preservative in accordance with manufacturer's instructions. Samples were analyzed for urine microalbumin (low levels compared to serum), CSF (cerebrospinal fluid or low-level) total protein, serum alkaline phosphatase, serum amylase, urine inorganic phosphate, serum lactate dehydrogenase (LD), serum alanine aminotransferase, serum aspartate aminotransferase (AST), serum creatinine kinase, serum lipase, and serum potassium, using a clinical analyzer (DXC 800). Neither the analyzer nor the analytical packs used on it were approved by the US Food and Drug Administration for use with oral fluid in any of its forms, so quantitative results were treated as relative rather than absolute. Samples also were analyzed for human immunoglobulins M, G, and A (IgM, IgG, and IgA) by a separate ELISA (enzyme-linked immunosubstrate assay) that was not a channel on the DXC 800.

From the above studies, inorganic phosphate, potassium, and total protein were eliminated from the list of potential SVT markers because detectable phosphate was found in one collector's buffer/preservative, potassium was found in another manufacturer's buffer/preservative, and a small amount of protein was found in one manufacturer's buffer/preservative. Of the remaining markers, only amylase, AST, LD, IgA, and albumin were deemed usable in neat oral fluid and all three buffer/preservatives. Table 6.1 summarizes the findings. Each rating is a qualitative composite of the authors' impression from

## Table 6.1 Decision Matrix for Oral Fluid SVT Markers

| Parameter | Amylase | AST | LD | IgA | Albumin |
|---|---|---|---|---|---|
| Overall rating | +8 | +3 | +7 | +4 or +7 | +8 |
| Analytical range[a] | +1 | 0 | +1 | +2 | +1 |
| Cost per test[b] | 0 | 0 | −1 | −2 | +2 |
| Routine measurement[c] | +2 | +1, decreasing signal rather than increasing | +2 | −2 if ELISA, +1 if nephelometric | +2 |
| Discriminatory power[d] | +2, more than ample analyte present | 0, questionable and needs further study | +1, ample analyte appears to be present in neat and diluted oral fluid in sufficient activity | +2, more than ample analyte appears to be present | +1, appears to be at the lower limit of sensitivity in the pre-study |
| Influence of biological variables[e] | 0, needs further study especially in salivary inflammation | 0, needs further study especially in liver disease | +1, housekeeping enzyme, ubiquitous, but needs further study | 0, may fluctuate with inflammation, needs further study | 0, one publication says fluctuates with dentation[6], needs further study |
| Transportability[f] | +1, stable enzyme; may fluctuate with mucin precipitation | 0, moderately stable enzyme, but still a protein | 0, stable enzyme, but still a protein | 0, stable enzyme, but still a protein | +1, stable but still a protein |
| Ability to be used in synthetic oral fluid[f] | +2 | +2 | +2 | +2 | 0 |
| Influenced by collection devices[g] | 0 | 0 | +1 | +2 | +2 |

## Key to Table 6.1

[a]Basis is a good match between the routine clinical range (vide infra) and the amount or activity of the analyte in both neat and 1:3 (Oral-EZE and Intercept) or 1:4 (Quantisal) dilutions.

[b]Cost per test was calculated using a discounted 2011 price list to client divided by 3 since a rough clinical formula is expendables × 3 = total cost including capital, expendables, labor, and overhead. The actual price for a given laboratory test will vary with the reagent manufacturer and volume discount from the manufacturer. The best cost per test, albumin

at $1.38, was +2. The most expensive test, IgA at $7.29, was a −2. Other test values were calculated as a point on a scale of $7.29−$1.38 in a fashion similar to the calculation of a Z-score in proficiency testing and rounded to the nearest whole number.

[c]Entries are based on whether offline (analysis separate from oral fluid drugs on an immunoassay analyzer) or online (analysis contemporaneously with oral fluid drugs on an immunoassay analyzer) and any other steps such as predilution.

[d]Discriminatory power is adequate separation between the lower limit of the oral fluid reference range in normal oral fluid and the substituted range (e.g. water or saline). Relative reportable ranges for analytes considered in the table above are

Amylase: 20−125 U/L
AST: 5−50 U/L
LD: 100−190 U/L
IgA: unknown at the time of publication
Albumin (microalbumin): 0.4−26.1 mg/L

[e]Entries are based on pre-study and clinical experience only. The effect of biological variables such as diurnal variation, state of hydration, psychosomatic influence, and disease states needs to be proven in an extended study.

[f]Entries indicate usefulness in proficiency testing programs where the matrix may be a synthetic that is not formulated with all of the components of authentic human oral fluid.

[g]Entries are based on serum, plasma, urine, and/or CSF stabilities from the LabCorp Online Manual (Beacon; https://www.labcorp.com/wps/portal/provider/testmenu, accessed 8/19/2011); Burtis CA, Ashwood ER, Bruns DE, eds. *Tietz Textbook of Clinical Chemistry and Molecular Diagnostics* (4th ed). St. Louis: Elsevier-Saunders; 2006; and the Quest Test Menu    (https://questdiagnostics.com/hcp/testmenu/jsp/showTestMenu.jsp? fn = 539.html&labCode = SKB, accessed 8/19/2011). Serum and body fluid amylase were maintained for 14 days at room temperature, urine amylase 14 days refrigerated. Serum AST (aspartate aminotransferase) is stable for 7 days according to the Laboratory Corporation of America Life Cycle Assessment (LCA); however, Panteghini, Bais, and van Solings in *Tietz* state "48 hours at 4°C." Serum LD is stable for 14 days at room temperature. IgA is stable for 3 days at room temperature according to the LCA Test Menu. Also, according to the LCA Test Menu, urine microalbumin is stable for 7 days at room temperature. Any SVT analyte chosen for further study also should be reviewed for transportability in oral fluid or buffer/preservative at room temperature and under any anticipated shipping conditions.

hands-on experience with the analytes as clinical chemistry tests and their application to oral fluid. If an analyte is assigned a positive integer, the authors' impression was that it found applicability for that category. In contrast, a negative integer meant that the analyte was either not suitable or, relative to other analytes, less suitable. Where a quantitative assignment such as cost per test could be made, the range is presented in the footnote for the category.

Initially, amylase and IgA appeared to be unusable due to high levels in human oral fluid. However, since it is a low or absent amount that would trigger further examination of a sample for dilution or substitution, these two markers should still be considered for SVT. So should uric acid, with a lower cutoff. It is notable that oral fluid IgA is a dimer, whereas the monomer is routinely measured in serum. Any validation of IgA as an SVT marker should include proof of acceptable cross-reactivity of the oral fluid dimeric form in a serum-based assay for IgA.

Zinc can have an adverse effect on common immunoassays for drugs and their metabolites.[7,8] So can surfactants such as soap.[9–11] White et al.[12] investigated the potential for oral cavity pharmaceuticals to be used as adulterants. The oral cavity rinse Biotène, Zicam spray, and the Zicam RapidMelts lozenge were examined as potential interferants in oral fluid drug testing. In the in vitro portion of the study, an 88% Zicam spray sample was formulated in synthetic oral fluid and spiked with $d$-amphetamine, $d$-methamphetamine, benzoylecgonine, morphine, phencyclidine (PCP), and 11-nor-$\Delta^9$-tetrahydrocannabinol. Testing by multiple immunoassays and chromatography–mass spectrometry was accomplished through the NLCP Oral Fluid Pilot PT Program. For one experiment in the in vivo segment, each of the two donors rinsed his mouth with a tablespoon of Biotène, expectorated the contents of his oral cavity to waste, and then collected a 50-mL neat oral fluid sample. In a second experiment in the in vivo segment, each donor held a RapidMelts lozenge in his oral cavity while collecting a 50-mL neat oral fluid sample.

In vitro, Biotène appeared to have an adverse influence on initial testing of oral fluid by immunoassay for parent THC, with both false negatives and false positives being produced. A similar but less severe influence was seen with initial testing by immunoassay for cocaine as the metabolite benzoylecgonine. Immunoassays such as those for amphetamine, methamphetamine, MDMA, opiates, 6-acetylmorphine, and PCP did not appear to be influenced by Biotène. Effects of Biotène on immunoassays in vivo were markedly attenuated, if present at all, relative to in vitro effects. In the in vivo experiments, spiked parent THC was markedly depressed (as much as 98% loss) relative to theoretical in both donors' oral fluid samples.

In vitro, Zicam spray adversely influenced initial testing of oral fluid by immunoassay primarily for parent THC, cocaine as benzoylecgonine, and opiates. Misapplication of the directed cutoff in the NLCP Oral Fluid Pilot PT Program also might affect interpretation of results. There was no significant in vivo effect of the lozenge relative to baseline. A notable observation: If oral fluid is collected and processed as the neat fluid, the pink color produced by the lozenge is easily recognizable visually, so an alert sample collector can thwart the use of this product to beat the system. Also, the pH of neat fluid with the lozenge present was almost 1 pH unit lower than that of the baseline sample but not abnormally low. For both potential adulterants, the in vitro study represented a worst-case scenario.

Although DNA (deoxyribonucleic acid) is used primarily to establish identity or nonidentity unequivocally, DNA also might be employed to establish that a sample is of human origin. Neat oral fluid is a rich source of nucleated buccal cells.

Fig. 6.1 shows partial collection of neat oral fluid. The sediment in the bottom of the tube is due to settling, not to centrifugation. It is a

*Figure 6.1 Partial (10 mL) neat oral fluid collection.*

mixture of precipitates, cells, and debris from the oral cavity. Fig. 6.2 is a 400 × magnification of a microdroplet of sediment that has been stained with methylene blue and cover-slipped. Nucleated cells are evident. The dark clumps in the photomicrograph are most likely bacteria that have been aggregated by IgA. Other debris were not identified.

White et al.[13] found that when neat oral fluid was extracted and subjected to analysis for short tandem repeats (STRs),[14,15] a full panel or 15 STRs plus amelogenin was obtained from all five donors. However, when the fluid was frozen for approximately 1 month and thawed, a complete panel of STRs plus amelogenin was obtained for only four of the five. When the same analyses were performed on the buffer/preservative and pads from a second-generation i2 Intercept, an Oral-EZE, and a second-generation Quantisal, all devices yielded a full panel of STRs plus amelogenin. However, after being stored for approximately a month, only the Quantisal yielded a full panel of STRs plus amelogenin for all five oral fluids.

At the time of this writing, no published articles on the presence or absence of intact cells in the three types of oral fluid buffer/preservative could be located. Likewise, no published articles on whether human DNA does or does not adsorb to any of the three types of collection pads could be extracted from the open literature or any proprietary publications. An oral fluid collector (ORAgene Dx OGD-610) designed specifically for harvesting human DNA for sequencing is commercially available.[16]

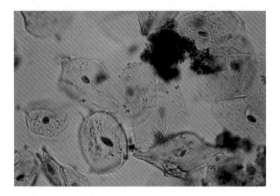

*Figure 6.2 Oral fluid, sedimented cells, and other debris (400×).*

## CONCLUSION

SVT is not as necessary for oral fluid drug testing as it is for urine drug testing. However, it still has a place for the reasons given at the beginning of the chapter. Should an oral fluid drug testing laboratory decide to employ SVT in their analytical profile, numerous methods exist or can be adapted from studies on the subject.

## REFERENCES

1. Mandatory Guidelines for Federal Workplace Drug Testing Programs, Substance Abuse and Mental Health Services, Department of Health and Human Services, 82 FR 7920-7970; 2017.

2. OraSure, Bethlehem, PA.

3. White RM, Hart ED. Unpublished results, Research Triangle Park, North Carolina: RTI International; 2012–17.

4. Crouch DJ, Day J, Baudys J, Fatah AA. Evaluation of saliva/oral fluid as an alternate drug testing specimen, Department of Justice Document Number, Award Number 94-IJ-R-004, 203569; February 2005.

5. Friesen L, Mitchell JM, White RM, Wong SH. Oral fluid specimen validity testing pre-study, RTI Project Proposal No. 0211848.001.002.00D.D02.110 Report; 2011.

6. Terrapon B, Mojon PH, Mensi N, Cimasoni G. Salivary albumin of edentulous patients. *Arch Oral Biol.* 1996;41(12):1183–1185.

7. Venkatratnam A, Lents NH. Zinc reduces the detection of cocaine, methamphetamine, and THC by ELISA urine testing. *J Anal Toxicol.* 2011;35:333–340.

8. Lin C-N, Strathman FG. Elevated urine zinc concentration reduces the detection of methamphetamine, cocaine, THC and opiates by EMIT. *J Anal Toxicol.* 2013;37(9):665–669.

9. Bronner W, Nyman P, von Minden D. Detectability of phencyclidine and 11-nor-$\Delta^9$-tetrahydrocannabinol-9-carboxylic acid in adulterated urine by radioimmunoassay and fluorescence polarization immunoassay. *J Anal Toxicol.* 1990;14:368–371.

10. Mikkelsen SL, Ash KO. Adulterants causing false negatives in illicit drug testing. *Clin Chem.* 1988;34(11):2333–2336.

11. Warner A. Interference of common household chemicals in immunoassay methods for drugs of abuse. *Clin Chem.* 1989;35(4):648–651.

12. White RM, Sutheimer C, Mitchell JM, Hart ED, Weber FX, Lodico C. The influence of zinc-containing over-the-counter products and an oral rinse on oral fluid drug testing. Submitted for publication.

13. White RM, Mitchell JM, Hart ED, et al. Assessment of the stability of DNA in specimens collected under conditions for drug testing—a pilot study. *Forensic Sci Int.* 2017;283:41–46.

14. Butler JM. *Advanced Topics in Forensic DNA Typing: Methodology.* Waltham, MA: Elsevier Academic Press; 2012.

15. Butler JM. Genetics and genomics of core short tandem repeat loci used in human identity testing. *J Forensic Sci.* 2006;51(2):253–265.

16. <Helix.com> Accessed 10.09.2017.

# CHAPTER 7

# Confirmatory Testing

## INTRODUCTION

In a 2009 review, Bosker and Huestis stated that between 2004 and 2008 there were reports of 71 oral fluid drug assays published using confirmatory or definitive techniques.[1] Since 2009, that number has increased dramatically because of the routine availability of liquid chromatography—mass spectrometry (LC-MS) systems. LC-MS techniques are beginning to dominate confirmatory testing in toxicology in general and oral fluid drug testing in particular because of the small sample volume available for analysis and the sensitivity required to detect low drug concentrations in oral fluid. There are several technological reasons for the shift towards LC-MS platforms in laboratory analysis, which include:

- Development of atmospheric pressure ionization (API) sources such as electrospray (ESI)[2] and atmospheric pressure chemical ionization (APCI)[3] applicable to small molecule analysis. Wang et al. compared three API configurations in the detection of 17 drugs in neat oral fluid and concluded that ion suppression for most of the analytes using ESI was lower than that of APCI and atmospheric pressure photoionization.[4]
- Replacement of traditional high-performance liquid chromatography (HPLC), which provides moderate separation and speed, with ultra-performance (UPLC) systems. UPLC instruments can operate with the increased back pressures generated when smaller bore, shorter chromatographic columns, and reduced particle sizes are incorporated into the analysis. These more efficient columns produce sharp, resolved analyte peaks, thereby improving sensitivity and reducing analysis time. Examples include the analysis of 29 drugs in oral fluid samples collected with the Saliva Sampler device in less than 15 min.[5]

Detection of Drugs and Their Metabolites in Oral Fluid. DOI: https://doi.org/10.1016/B978-0-12-814595-1.00007-6

- Availability of new technology combines the principles of turbulence, diffusion, and chemistry to eliminate matrix interferences from specimens. This allows direct injection of biological samples into the mass spectrometer.
  Note: While this advance in technology is helpful for biological matrices that can be diluted and injected, many oral fluid collection devices include a transportation buffer that contains surfactants, preservatives, and antimicrobial agents (see Chapter 3). These additives will affect the chromatography produced by the LC-MS/MS system and may cause significant internal issues with the mass spectrometer.
- Multiplexing on LC-MS platforms. This allows the mobile phases from multiple LC columns to be directed to a single mass spectrometer so that the number of analyses that can be performed in a given time is increased.

These advancements can be incorporated and connected to different types of mass spectrometers—e.g., single quadrupole, triple quadrupole (QQQ), time-of-flight (ToF), linear ion trap, quadrupole time-of-flight (Q-ToF), and even triple quadrupole scan modes combined with linear ion trap operations (Q-Trap) to produce enhanced specificity and sensitivity in a bench-top format.

LC-MS systems are straightforward to use, provide fast run times, have excellent specificity, and generally do not require drug derivatization for improved chromatography and sensitivity, in contrast to gas chromatography—mass spectrometry (GC-MS). Even though LC-MS instruments are becoming more popular, confirmatory techniques for drugs in oral fluid using GC-MS following immunoassay screening are still used.[6–8]

## SEPARATION

In GC-MS the carrier gas (usually helium or hydrogen) moves the compounds through the analytical column to the mass spectrometer for detection; in LC-MS it is the liquid mobile phase that carries the analytes along the column and into the ionization source. Mobile phases can be run isocratically or in gradient fashion where two or more liquids mix in different percentages throughout the run. The advantage of gradient runs is improved chromatography and speed; the disadvantage is the potential instability of the system prior to the next injection. As with all assays that are required to be sensitive, robust, and reproducible, all chemicals, salts, and solvents should be of high purity (typical reagents are ammonium formate, ammonium acetate, formic acid, methanol, and acetonitrile).

## METHOD VALIDATION

There are different criteria for determining positivity of specimens in chromatography and spectrometry. Validation parameters for the determination of drugs in oral fluid are the same as those necessary for other analytical methods in toxicology. These should include linearity, limits of detection and quantitation, precision, accuracy, and specificity; additional parameters for LC-MS/MS assays include measurement of ion suppression, matrix effects, and process efficiency. Unique variables in the analysis of oral fluid are drug recovery from a collection device (if used) and drug stability in transportation buffers and during storage. In GC-MS, data acquisition is generally by selected ion monitoring; in LC-MS/MS, selected reaction monitoring and multiple reaction monitoring are dominant techniques. The importance of monitoring and using more than one transition for drug identification in LC-MS/MS has been published, especially when groups of drugs with the same molecular weights are being monitored in the same assay.[9,10] Monitoring of three transitions to produce two ion ratios is recommended for urine testing, but for oral fluid analysis where the drug concentration is likely to be significantly lower, it is not always possible to achieve three transitions and maintain the sensitivity required for detection. However, it is critical to monitor at least two specific transitions (one ratio) to minimize the potential for misidentification of compounds with similar fragmentation patterns. Professional guidance is available on acceptable validation of methods for drug analysis in oral fluid.[11]

While the majority of published methods use QQQ instruments, linear or three-dimensional ion trap instruments as well as ToF and Q-ToF mass analyzers have also been used to analyze drugs in oral fluid.[3,12] To avoid overloading the ion trap, di Rago et al. used a novel approach, carbon-13 isotopes of the analytes, to reduce detector saturation.[3] ToF and Q-ToF instruments are increasingly being implemented where the drug or drug class being searched for is unknown (e.g., in forensic analysis); if the procedure is designed as a confirmatory method, this would not be the case, because the definitive technique is confirming the presence or absence of a presumptive initial test (e.g., immunoassay).

Confirmation of drugs and metabolites in oral fluid usually involves extraction of the analytes from the fluid itself or oral fluid/buffer mix from collection devices. Even when neat oral fluid is collected, there are still sample preparation steps remaining before injection into an instrument, e.g., centrifugation and precipitation.[13] Dilution with buffer, supported liquid extraction (SLE), and elution with ethyl acetate—hexane have also been reported.[14] When collection devices with no buffer are used, the drugs are eluted from the pad or device using an organic solvent, which may be directly injected into the instrument.

The most common method of drug extraction from oral fluid samples collected in buffer used to be liquid—liquid extraction (LLE),[15] but now SLE[16] and solid-phase extraction (SPE) are routinely used.[17,18] SPE concentrates the analytes of interest onto a sorbent bed and removes other matrix interferences; this helps with achieving the sensitivity required for drug identification in a small volume of biological sample and decreases ion suppression issues in LC-MS/MS.[19] In 2012, da Fonseca et al. developed a procedure for the identification of biomarkers for tobacco smoke exposure using SPE with GC-MS/MS; it required only 0.2 mL of oral fluid.[20]

Other extraction techniques have been reported, including supercritical fluid extraction (not widely used) and a form of miniaturized SPE known as micro-extraction by packed sorbent (MEPS) which allows limited volume biological samples to be extracted and the analytes concentrated for analysis.[21,22]

## SENSITIVITY

Proposed cutoff concentrations for several drug classes in oral fluid for workplace testing have been published (Table 7.1). These levels can be achieved with standard GC-MS and LC-MS instruments.

Test concentrations for other applications (e.g., medication compliance, criminal justice, sports testing) may be different.

| Table 7.1 Cutoff Concentrations for Confirmatory Tests in Oral Fluid (ng/mL) | | | |
|---|---|---|---|
| | Geographical Area and Professional Organization | | |
| Target Compound (not all countries include all the drugs in a class) | United States (SAMHSA)[23] | Europe (EWDTS)[11] | Australia (AS4760:2017 revision)[24] |
| amphetamine, methamphetamine, MDMA, MDA | 15 | 15 | 25 |
| cannabis (THC) | 2 | 2 | 5 |
| cocaine, benzoylecgonine, ecgonine methyl ester | 8 | 8 | 25 |
| codeine, morphine, hydrocodone, hydromorphone, | 15 | 15 | 25 |
| dihydrocodeine | | 15 | |
| oxycodone, oxymorphone | 15 | | 20 |
| 6-acetylmorphine | 2 | 2 | 10 |
| methadone | | 20 | |
| benzodiazepines (up to 18 listed in EWDTS) | | 3 (temazepam = 10) | 10 |
| phencyclidine | 2 | | |

## MAJOR DRUG CLASSES

### Amphetamines

Basic drugs, such as amphetamines, diffuse into saliva from the blood relatively easily and are present in measurable concentrations following use. There are many publications on their extraction and analysis, predominantly using LLE, solid-phase procedures, GC-MS,[7,19,25] and LC-MS/MS.[26] Methamphetamine and its metabolite amphetamine are easily detected by routine techniques; other sympathomimetic amines such as 3,4-methylendioxymethylamphetamine (MDMA, called "ecstasy") and 3,4-methylendioxyamphetamine are also detected in high concentrations. In 2016, Poetzsch et al.[27] described a new technique involving matrix-assisted laser desorption/ionization-triple quadrupole–tandem mass spectrometry (MALDI-QQQ-MS/MS) which provides an ultrafast high-throughput platform. Oral fluid samples collected with the Quantisal device were buffered and subjected to liquid–liquid extraction (ButOAc/EtOAc, 1:1). Only 1 μL of the extract was necessary for analysis. The extract and the MALDI matrix (alpha-cyano-4-hydroxycinnamic acid) were spotted onto the MALDI plate and analyzed. Because the MALDI technique has no chromatographic separation, the analysis was done within 10 s. The authors analyzed 250 oral fluid samples for MDMA within an hour; results compared well with routine LC-MS/MS.

## Cocaine

Following cocaine administration, the parent drug itself is the predominant compound detected in oral fluid, appearing within minutes regardless of the route of administration. Using GC-MS for confirmation, Scheidweiler et al. reported that in oral fluid collected via expectoration after subcutaneous administration, cocaine was detected within 5–30 min (depending on dose) and the half-life was 1.1–3.8 h. The metabolites benzoylecgonine (BZE) and ecgonine methyl ester had longer half-lives, up to 15 h. Comparison of these results with those produced following Salivette collections did not show any significant differences.[28]

Ellefson et al. reported oral fluid concentrations of cocaine and BZE from different collection devices, Oral-Eze and StatSure Saliva Sampler, in 10 adults after intravenous administration of cocaine (25 mg). Samples were collected at various time points from 1 h before administration to 69 h afterward. The authors reported variability between subjects, but cocaine was detected in the oral fluid within 10 min. The median concentration detected using the Oral-Eze was 932 ng/mL of cocaine and 248 ng/mL of BZE; using the StatSure, 732 ng/mL of cocaine and 360 ng/mL of BZE. Detection time with the same limit of quantitation was longer with the Oral-Eze. The authors noted that the cocaine half-life was substantially shorter in the StatSure samples, a difference they attribute to stabilizing buffers in the devices or decreased drug recovery from the collection pads.[18]

Three publications report the use of LC-MS to detect cocaine and its metabolites in oral fluid. Ares et al. used MEPS followed by UPLC-MS/MS to identify cocaine and two of its metabolites, along with 11 synthetic cathinones, 6 opiates, and scopolamine, in less than 3 min.[21] Fiorentin et al. used a single quadrupole LC-MS for simultaneously analyzing cocaine, BZE, cocaethylene, anhydroecgonine methyl ester, and anhydroecgonine in oral fluid, urine, and plasma.[29] Bombana et al. used a UPLC-MS/MS method for oral fluid from truck drivers in Brazil; cocaine was the most prevalent drug detected, even more than amphetamines and $\Delta^9$-tetrahydrocannabinol (THC).[30]

## Cannabis

The most widely used illicit drug in the world is marijuana, whose main psychoactive component is THC. While THC itself is easily detected in oral fluid collected by various devices using GC-MS[31,32] or LC-MS/MS,[33,34] its metabolite 11-nor-$\Delta^9$-tetrahydrocannabinol-9-carboxylic acid (THC-COOH) is present in such low concentration, even in the saliva of frequent marijuana users, that more sensitive confirmation techniques such as two-dimensional GC-MS[35,36] and GC-MS/MS were developed.[37] However, some procedures that analyze simultaneously for THC and THC-COOH do not reach the requisite sensitivity to detect THC-COOH in marijuana users.[38]

THC can also be detected in oral fluid after passive exposure to marijuana,[39,40] so these methods to detect THC-COOH, the presence of which appears to differentiate between use and exposure, are important in drug testing applications.

The disposition of cannabis in oral fluid collected with the Quantisal device after smoking, vaping, and edible intake was reported by Swortwood et al.[41] Briefly, the specimens were hydrolyzed with base, neutralized, extracted by cation exchange SPE, and analyzed by LC-MS/MS using APCI. The assay detected THC, 11-hydroxy-THC (11-OH-THC), $\Delta^9$-tetrahydrocannabavarin, cannabidiol, cannabigerol, and THC-COOH. The limits of quantitation were 0.2 ng/mL for all except THC-COOH, which was 15 pg/mL. Few differences were observed between smoking and vaping, with THC peaking almost immediately; the time to maximum concentration of THC after oral consumption (edibles) was approximately 24 min.

## Benzodiazepines

Benzodiazepines are an important group in pain management, medical compliance monitoring, and driving under the influence of drugs cases. They are included in workplace testing guidelines in Europe and Australia, but not in the federal proposals in the United States. Xanax (alprazolam), Klonopin (clonazepam), and Valium (diazepam) are widely prescribed. They have a high degree of protein binding and low saliva—plasma ratios and do not accumulate well in oral fluid, which makes detection difficult. However, LC-MS techniques lend themselves well to this group because no derivatization is required, multiple drugs with similar chemical properties can be analyzed under the same analytical conditions, and the instruments are adequately sensitive to achieve the drug concentrations required. Laboratory methods using direct injection of collected oral fluid into the LC-MS/MS have been reported.[42] Other methods using QQQ detectors, linear ion traps, and ToF instruments have been published for this drug class.[15,43,44]

Generally, parent drugs are detected in higher concentration in oral fluid than the metabolites. An interesting exception is the metabolite of clonazepam, 7-aminoclonazepam. In 2016, Melanson et al. reported for the first time that analysis of 100 oral fluid specimens collected with the Intercept device and presumptively positive for clonazepam and its metabolite showed a higher percentage positive for the metabolite (91%) than the parent (44%); all samples were analyzed by LC-MS/MS. The concentrations of 7-aminoclonazepam were approximately 2.4-fold higher than clonazepam itself.[45] In a separate study, Vindenes et al. (2016) analyzed over 1000 oral fluid samples positive for clonazepam, nitrazepam, or flunitrazepam; they confirmed that the 7-amino metabolites were more likely to be detected than the parent.[46]

## Opioids

The opioid group includes compounds from federal workplace testing programs—codeine, morphine, and 6-acetylmorphine (an active metabolite of heroin)—as well as several widely prescribed medications for pain management or heroin substitution programs, such as hydrocodone, hydromorphone, oxycodone, oxymorphone, buprenorphine, methadone, fentanyl, and tramadol. Most of these drugs have saliva−plasma ratios >1, so they accumulate well in oral fluid and can be easily detected via routine GC-MS[47,48] or LC-MS.[49−54] In 2011, Tuyay et al. reported the disposition of opioids in oral fluid collected with the Quantisal device. The article specifically focused on the importance of chromatography and mass spectral transitions in LC-MS/MS.[9] After intake of codeine, the main metabolite in urine is morphine but in oral fluid it is norcodeine. The same is true for oxycodone, where noroxycodone is the major metabolite, not oxymorphone,[14] and for hydrocodone, where norhydrocodone, not hydromorphone, is the main metabolite in oral fluid.[51] Problems arise in LC-MS/MS assays when compounds have the same molecular weight and potentially the same fragmentation patterns, and so are not separated chromatographically.[10] Fig. 7.1 shows the molecular structures of some opioids and their demethylated metabolites, and Fig. 7.2 shows different

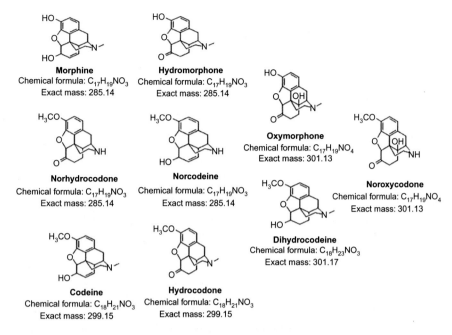

Figure 7.1 Molecular structures of opioids and their demethylated metabolites. Molecular weight m/z 285: morphine, hydromorphone, norhydrocodone, norcodeine. Molecular weight m/z 299: codeine, hydrocodone. Molecular weight m/z 301: oxymorphone, noroxycodone, dihydrocodeine. Reproduced with permission from Tuyay J, Coulter C, Rodrigues W, Moore C. Disposition of opioids in oral fluid: importance of chromatography and mass spectral transitions in LC-MS/MS. Drug Test Anal. 2012;4(6):395−401.

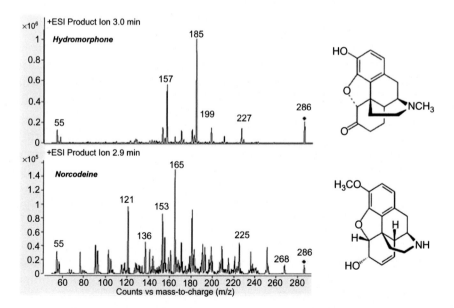

Figure 7.2 Product ion full scan obtained by collisions of M + H + ions of hydromorphone and norcodeine. Reproduced with permission from Tuyay J, Coulter C, Rodrigues W, Moore C. Disposition of opioids in oral fluid: importance of chromatography and mass spectral transitions in LC-MS/MS. Drug Test Anal. 2012;4(6):395−401.

fragmentation for two opioids with the same molecular weight. Robust chromatography combined with multiple transition monitoring is essential to differentiate between structurally related compounds.

## Phencyclidine

Phencyclidine (PCP) maintains its place on the US Substance Abuse and Mental Health Services Administration (SAMHSA) list of drugs that may be tested for in federal workplace settings. In 2008, two research groups reported the quantitation of PCP in oral fluid in specimens collected with the Quantisal device. Coulter et al. used an Agilent 1200 series connected to a 6410 LC-MS/MS instrument and established a limit of detection of 5 ng/mL.[55] Kala et al. used a Shimadzu HPLC connected to an Applied Biosystems 3200 QTRAP LC-MS/MS and established a limit of detection of 2 ng/mL.[26] Both groups used deuterated internal standards to minimize ion suppression and monitored two multiple reaction mode transitions in positive ESI mode.

## NOVEL PSYCHOACTIVE SUBSTANCES

GC-MS and LC-MS/MS can be used to analyze novel psychoactive substances (NPS), including synthetic cathinones and synthetic cannabinoids, in oral fluid.[17,56,57] As more potent designer drugs are synthesized—piperazines, tryptamines, designer opioids, etc.—the concentrations detected in oral fluid are low and the limited availability of reference standards inhibits routine quantitative analysis. Most recent methods can detect a wide range of these emerging drugs in one assay, often 20–30 of them, at low limits of detection. However, the range of drugs in the test profile is often geographically driven. To evaluate the impact of NPS on recreational drug toxicity in Oslo, Norway, oral fluid and blood samples were analyzed by UPLC-MS/MS.[58] Most often detected were clonazepam, amphetamines, and heroin, but also 4-methylamphetamine, dimethyltryptamine, methylone, N,N-dimethyl-3,4-methylenedioxyamphetamine, JWH-210, AM-2201, and 1-(benzofuran-5-yl)-N-ethylpropan-2-amine (5-EAPB) were detected. Overall, NPS were detected in 8% of cases.

Recent reviews of the analytical methods for the detection of synthetic cannabinoids suggest that nontargeted high-resolution mass spectrometry screening, which is more flexible and permits retrospective data analysis, is a promising strategy to maintain relevant assays.[59]

More detailed information on various LC-MS techniques used for drug detection in oral fluid can be found in review articles and vendor parameter libraries.[60,61]

## DERIVATIZATION IN LC-MS

One of the advantages of LC-MS over GC-MS is that derivatization is not necessary to achieve the required detection limits for drugs in oral fluid. However, there are situations which require the use of derivatives.

### Chiral Separation of Amphetamine and Methamphetamine

In workplace testing programs, it is often necessary to differentiate between the isomers of amphetamine and methamphetamine (*l*-AMP, *l*-MAMP, *d*-AMP, *d*-MAMP) or racemic mixtures (*d,l*-AMP, *d,l*-MAMP). Chiral derivatization has been reported using Marfey's reagent as the derivatizing agent; ESI in negative mode was used for ionization. Even though the m/z transitions are the same, the enantiomers are separated chromatographically on a standard LC-MS column. The separation of the enantiomers in oral fluid collected with the Quantisal and the Oral-Eze was reported; the procedure had a limit of detection of 0.5 ng/mL (Fig. 7.3).[62]

The prescription medication Adderall is a mixture of neutral salts of *d*-amphetamine sulfate, amphetamine sulfate, the *d*-isomer of amphetamine saccharate, and *d,l*-amphetamine aspartate. For each tablet, the combination of salts and isomers results in a 3:1 ratio of *d:l*-amphetamine. Specimens from a volunteer taking prescription Adderall showed mean *d*-AMP and *l*-AMP concentrations of 133 and 44 ng/mL, respectively (medians 125 and 44 ng/mL, respectively) with an average ratio of 2.9:1 (median 2.8:1). Even though the mean elimination half-life ($t_{1/2}$) for *d*-AMP is shorter than the $t_{1/2}$ of the *l*-isomer (9.7−11 h vs 11.5−13.8 h), the measured ratio held throughout the sample collection times (over 31 h). An approximate 3:1 ratio of *d*-AMP to *l*-AMP in oral fluid specimens, with no methamphetamine present, is potentially indicative of Adderall use.[63]

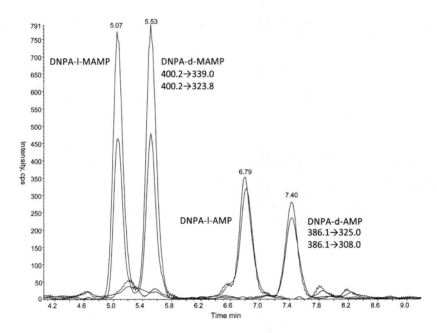

*Figure 7.3 Derivatized methamphetamine (dinitrophenylalaninamide-methamphetamine, DNPA-MAMP) and derivatized amphetamine (dinitrophenylalaninamide-amphetamine, DNPA-AMP) extracted ion chromatograms at the limit of quantitation in Quantisal.* Reproduced with permission from Newmeyer MN, Concheiro M, Huestis MA. Rapid quantitative chiral amphetamines liquid chromatography-tandem mass spectrometry: method in plasma and oral fluid with a cost-effective chiral derivatizing reagent. J Chromatogr A. 2014;1358:68−74. doi: 10.1016/j.chroma.2014.06.096.

## Improved Sensitivity for THC-COOH in Oral Fluid

As discussed previously, the determination of THC in oral fluid using either GC-MS or LC-MS is straightforward. However, the concentration of the metabolite THC-COOH is approximately 1000 times lower. The detection of THC-COOH is important to differentiate potential passive exposure to marijuana from actual intake.[39,40] The requisite sensitivity was achieved on a bench-top triple quadruple LC-MS/MS following a rapid simple derivatization procedure which did not affect the response of THC, analyzed simultaneously. Briefly, triphenylphosphine (10 mM) in acetonitrile was added to the dried oral fluid extract; 2,2'-dithiodipyridine (10 mM), DPDS in acetonitrile, and 2-picolylamine (10 μg) were successively added, and the mixture was heated at 60°C for 15 min. The mixture was evaporated to dryness and reconstituted for LC-MS/MS analysis. Following derivatization, the THC-COOH two 2-picolylamine transitions were 435.3 > 327 and 435.3 > 299.[64] Fig. 7.4 shows the fragmentation pattern for the derivatized THC-COOH.

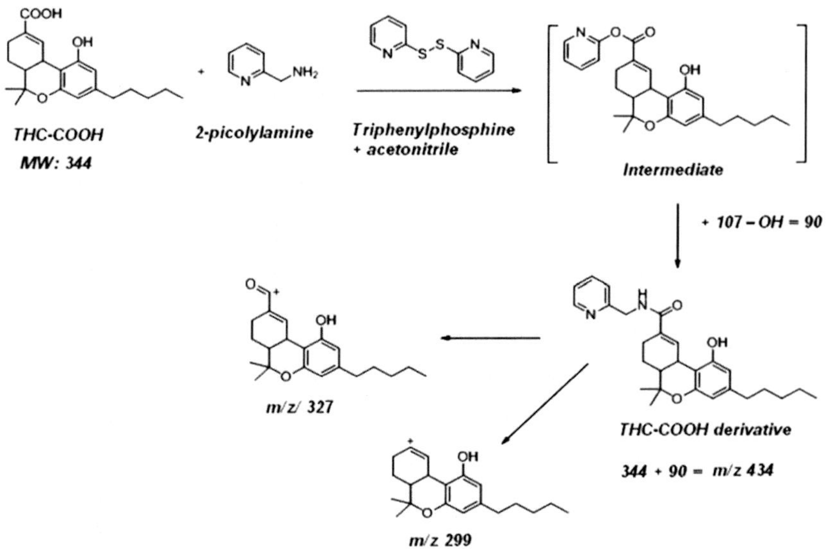

*Figure 7.4 Fragmentation pattern for derivatized THC-COOH.* Reproduced with permission from Coulter C, Garnier M, Moore C. Analysis of tetrahydrocannabinol and its metabolite, 11-nor-d9-tetrahydrocannabinol-9-carboxylic acid, in oral fluid using liquid chromatography with tandem mass spectrometry. J Anal Toxicol. 2012;36(6):413−7.

## Improved Sensitivity for 11-OH-THC in Oral Fluid

The active metabolite 11-OH-THC is also indicative of cannabis use as opposed to possible drug exposure when detected in oral fluid. However, the concentration even in frequent users is extremely low. To improve sensitivity, Thieme et al. reported significant enhancement of 11-OH-THC detection by formation of picolinic acid esters and analysis by multi-stage LC-MS/MS/MS (MS3). The derivatization improved the detection limits by a factor of 100 over underivatized extracts.[65]

## SUMMARY

- For quantitative analysis, the dilution factor from collection devices that incorporate transportation buffers must be taken into account for the calibrators and controls used in the assay.
- GC-MS instruments are relatively inexpensive, in routine use, robust, and easy to operate. To achieve the requisite sensitivity for detection of drugs in oral fluid, derivatization is generally necessary, which is an additional stage in the laboratory process and may involve exposure to hazardous chemicals.

- LC-MS instruments are becoming less expensive and more routinely implemented. Difficulties with analysis are predominantly related to ion suppression, failure to incorporate matched matrices or deuterated internal standards, and the potential for identical transitions, particularly from structurally similar drugs or metabolites.[9,10,66] Direct or dilute injection of biological samples such as urine is possible, but for oral fluid drugs should be efficiently extracted from transportation buffers to avoid injection of stabilizers and surfactants directly into instruments.
- More advanced mass spectrometers such as high-resolution instruments are useful to detect a wide range of compounds, especially those with no commercially available drug standards (e.g., synthetic cannabinoids and cathinones). However, the instruments are more costly and the expertise required for operation is of a higher level than needed to run bench-top LC-MS/MS systems.

## REFERENCES

1. Bosker WM, Huestis MA. Oral fluid testing for drugs of abuse. *Clin Chem.* 2009;55 (11):1910−1931.

2. Dams R, Choo RE, Lambert WE, Jones H, Huestis MA. Oral fluid as an alternative matrix to monitor opiate and cocaine use in substance-abuse treatment patients. *Drug Alcohol Depend.* 2007;87(2−3):258−267.

3. Di Rago M, Chu M, Rodda LN, Jenkins E, Kotsos A, Gerostamoulos D. Ultra-rapid targeted analysis of 40 drugs of abuse in oral fluid by LC-MS/MS using carbon-13 isotopes of methamphetamine and MDMA to reduce detector saturation. *Anal Bioanal Chem.* 2016;408 (14):3737−3749. Available from: https://doi.org/10.1007/s00216-016-9458-3.

4. Wang IT, Feng YT, Chen CY. Determination of 17 illicit drugs in oral fluid using isotope dilution ultra-high performance liquid chromatography/tandem mass spectrometry with three atmospheric pressure ionizations. *J Chromatogr B Analyt Technol Biomed Life Sci.* 2010;878 (30):3095−3105. Available from: https://doi.org/10.1016/j.jchromb.2010.09.014.

5. Badawi N, Simonsen KW, Steentoft A, Bernhoft IM, Linnet K. Simultaneous screening and quantification of 29 drugs of abuse in oral fluid by solid-phase extraction and ultra-performance LC-MS/MS. *Clin Chem.* 2009;55(11):2004−2018.

6. Langel K, Gunnar T, Ariniemi K, et al. A validated method for the detection and quantitation of 50 drugs of abuse and medicinal drugs in oral fluid by gas chromatography-mass spectrometry. *J Chromatogr B Analyt Technol Biomed Life Sci.* 2011;879(13−14):859−870. Available from: https://doi.org/10.1016/j.jchromb.2011.02.027.

7. Bahmanabadi L, Akhgari M, Jokar F, Sadeghi HB. Quantitative determination of methamphetamine in oral fluid by liquid-liquid extraction and gas chromatography/mass spectrometry. *Hum Exp Toxicol.* 2017;36(2):195−202. Available from: https://doi.org/10.1177/0960327116638728.

8. Cohier C, Mégarbane B, Roussel O. Illicit drugs in oral fluid: evaluation of two collection devices. *J Anal Toxicol.* 2017;41(1):71−76. Available from: https://doi.org/10.1093/jat/bkw100.

9. Tuyay J, Coulter C, Rodrigues W, Moore C. Disposition of opioids in oral fluid: importance of chromatography and mass spectral transitions in LC-MS/MS. *Drug Test Anal.* 2012;4 (6):395−401.

10. Fox EJ, Twigger S, Allen KR. Criteria for opiate identification using liquid chromatography linked to tandem mass spectrometry: problems in routine practice. *Ann Clin Biochem.* 2009;46:50−57.

11. Brcak M, Beck O, Bosch T, et al. European guidelines for workplace drug testing in oral fluid. *Drug Test Anal.* 2017:1−14. Available from: https://doi.org/10.1002/dta.2229.

12. Clauwaert K, Decaestecker T, Mortier K, et al. The determination of cocaine, benzoylec-gonine and cocaethylene in small-volume oral fluid samples by liquid chromatography-quadrupole-time-of-flight mass spectrometry. *J Anal Toxicol.* 2004;28:655−659.

13. Liu HC, Lee HT, Hsu YC, et al. Direct injection LC-MS-MS analysis of opiates, metham-phetamine, buprenorphine, methadone and their metabolites in oral fluid from substitution therapy patients. *J Anal Toxicol.* 2015;39(6):472−480. Available from: https://doi.org/10.1093/jat/bkv041.

14. Cone EJ, DePriest AZ, Heltsley R, et al. Prescription opioids. III. Disposition of oxycodone in oral fluid and blood following controlled single-dose administration. *J Anal Toxicol.* 2015;39(3):192−202. Available from: https://doi.org/10.1093/jat/bku176.

15. Jang M, Chang H, Yang W, et al. Development of an LC-MS/MS method for the simulta-neous determination of 25 benzodiazepines and zolpidem in oral fluid and its application to authentic samples from regular drug users. *J Pharm Biomed Anal.* 2013;74:213−222. Available from: https://doi.org/10.1016/j.jpba.2012.11.002.

16. Rositano J, Harpas P, Kostakis C, et al. Supported liquid extraction (SLE) for the analysis of methylamphetamine, methylenedioxymethylamphetamine and delta-9-tetrahydrocannabinol in oral fluid and blood of drivers. *Forens Sci Int.* 2016;265:125−130. Available from: https://doi.org/10.1016/j.forsciint.2016.01.017.

17. Miller B, Kim J, Concheiro M. Stability of synthetic cathinones in oral fluid samples. *Forens Sci Int.* 2017;274:13−21. Available from: https://doi.org/10.1016/j.forsciint.2016.11.034.

18. Ellefsen KN, Concheiro M, Pirard S, Gorelick DA, Huestis MA. Oral fluid cocaine and ben-zoylecgonine concentrations following controlled intravenous cocaine administration. *Forens Sci Int.* 2016;260:95−101. Available from: https://doi.org/10.1016/j.forsciint.2016.01.013.

19. Moore C, Coulter C, Crompton K. Achieving proposed Federal concentrations using reduced specimen volume for the extraction of amphetamines from oral fluid. *J Anal Toxicol.* 2007;31(8):442−446.

20. da Fonseca BM, Moreno IE, Magalhães AR, et al. Determination of biomarkers of tobacco smoke exposure in oral fluid using solid-phase extraction and gas chromatography-tandem mass spectrometry. *J Chromatogr B Analyt Technol Biomed Life Sci.* 2012;15 (889−890):116−122. Available from: https://doi.org/10.1016/j.jchromb.2012.02.006.

21. Ares AM, Fernández P, Regenjo M, et al. A fast bioanalytical method based on micro-extraction by packed sorbent and UPLC-MS/MS for determining new psychoactive sub-stances in oral fluid. *Talanta.* 2017;174:454−461. Available from: https://doi.org/10.1016/j.talanta.2017.06.022.

22. Montesano C, Simeoni MC, Curini R, et al. Determination of illicit drugs and metabolites in oral fluid by microextraction on packed sorbent coupled with LC-MS/MS. *Anal Bioanal Chem.* 2015;407(13):3647−3658. Available from: https://doi.org/10.1007/s00216-015-8583-8.

23. SAMHSA, May 15, 2015 Federal Register, 80 FR 28053, Oral Fluid Mandatory Guidelines; US Federal Government, Washington DC.

24. AS 4760:2006. *Procedures for Specimen Collection and the Detection and Quantitation of Drugs in Oral Fluid.* Standards Australia; 2006.

25. Mohamed K. One-step derivatization-extraction method for rapid analysis of eleven amphetamines and cathinones in oral fluid by GC/MS. *J Anal Toxicol.* 2017;41(7):639−645.

26. Kala SV, Harris SE, Freijo TD, Gerlich S. Validation of analysis of amphetamines, opiates, phencyclidine, cocaine, and benzoylecgonine in oral fluids by liquid chromatography-tandem mass spectrometry. *J Anal Toxicol.* 2008;32(8):605−611.

27. Poetzsch M, Steuer AE, Hysek CM, Liechti ME, Kraemer T. Development of a high-speed MALDI-triple quadrupole mass spectrometric method for the determination of 3,4-methylenedioxymethamphetamine (MDMA) in oral fluid. *Drug Test Anal.* 2016;8(2):235−240. Available from: https://doi.org/10.1002/dta.1810.

28. Scheidweiler KB, Spargo EA, Kelly TL, Cone EJ, Barnes AJ, Huestis MA. Pharmacokinetics of cocaine and metabolites in human oral fluid and correlation with plasma concentrations after controlled administration. *Ther Drug Monit.* 2010;32 (5):628−637. Available from: https://doi.org/10.1097/FTD.0b013e3181f2b729.

29. Fiorentin TR, D'Avila FB, Comiran E, et al. Simultaneous determination of cocaine/crack and its metabolites in oral fluid, urine and plasma by liquid chromatography-mass spectrometry and its application in drug users. *J Pharmacol Toxicol Methods.* 2017;86:60−66. Available from: https://doi.org/10.1016/j.vascn.2017.04.003.

30. Bombana HS, Gjerde H, Dos Santos MF, et al. Prevalence of drugs in oral fluid from truck drivers in Brazilian highways. *Forensic Sci Int.* 2017;273:140−143. Available from: https://doi.org/10.1016/j.forsciint.2017.02.023.

31. Yonamine M, Sanches LR, Paranhos BA, de Almeida RM, Andreuccetti G, Leyton V. Detecting alcohol and illicit drugs in oral fluid samples collected from truck drivers in the state of São Paulo, Brazil. *Traffic Inj Prev.* 2013;14(2):127−131. Available from: https://doi.org/10.1080/15389588.2012.696222.

32. Moore C, Vincent M, Rana S, Coulter C, Agrawal A, Soares J. Stability of $\Delta^9$-tetrahydrocannabinol (THC) in oral fluid using the Quantisal™ collection device. *Forensic Sci Int.* 2006;164(2−3):126−130.

33. Wille SM, Di Fazio V, Ramirez-Fernandez Mdel M, Kummer N, Samyn N. Driving under the influence of cannabis: pitfalls, validation, and quality control of a UPLC-MS/MS method for the quantification of tetrahydrocannabinol in oral fluid collected with StatSure, Quantisal, or Certus collector. *Ther Drug Monit.* 2013;35(1):101−111.

34. Samano KL, Anne L, Johnson T, Tang K, Sample RH. Recovery and stability of Δ9-tetrahydrocannabinol using the Oral-Eze® oral fluid collection system and Intercept® oral specimen collection device. *J Anal Toxicol.* 2015;39(8):648−654. Available from: https://doi.org/10.1093/jat/bkv093.

35. Moore C, Coulter C, Rana S, Vincent M, Soares J. Analytical procedure for the determination of the marijuana metabolite, 11-nor-$\Delta^9$-tetra-hydrocannabinol-9-carboxylic acid (THCA), in oral fluid specimens. *J Anal Toxicol.* 2006;30(7):409−412.

36. Newmeyer MN, Desrosiers NA, Lee D, Mendu DR, Barnes AJ, Gorelick DA, et al. Cannabinoid disposition in oral fluid after controlled cannabis smoking in frequent and occasional smokers. *Drug Test Anal.* 2014;6(10):1002−1010. Available from: https://doi.org/10.1002/dta.1632.

37. Day D, Kuntz DJ, Feldman M, et al. Detection of THCA in oral fluid by GC-MS-MS. *J Anal Toxicol.* 2006;30(7):645−650.

38. Bylda C, Leinenbach A, Thiele R, Kobold U, Volmer DA. Development of an electrospray LC-MS/MS method for quantification of D9-tetrahydrocannabinol and its main metabolite in oral fluid. *Drug Test Anal.* 2012;4(7−8):668−674.

39. Moore C, Coulter C, Uges D, et al. Cannabinoids in oral fluid following passive exposure to marijuana smoke. *Forensic Sci Int.* 2011;212(1−3):227−230.

40. Cone EJ, Bigelow GE, Herrmann ES, et al. Nonsmoker exposure to secondhand cannabis smoke. III. Oral fluid and blood drug concentrations and corresponding subjective effects. *J Anal Toxicol*. 2015;39(7):497−509. Available from: https://doi.org/10.1093/jat/bkv070.

41. Swortwood MJ, Newmeyer MN, Andersson M, Abulseoud OA, Scheidweiler KB, Huestis MA. Cannabinoid disposition in oral fluid after controlled smoked, vaporized, and oral cannabis administration. *Drug Test Anal*. 2017;9(6):905−915. Available from: https://doi.org/10.1002/dta.2092.

42. Nordal K, Øiestad EL, Enger A, Christophersen AS, Vindenes V. Detection times of diazepam, clonazepam, and alprazolam in oral fluid collected from patients admitted to detoxification, after High and repeated drug intake. *Ther Drug Monit*. 2015;37(4):451−460. Available from: https://doi.org/10.1097/FTD.0000000000000174.

43. Moore C, Coulter C, Crompton K, Zumwalt M. Determination of benzodiazepines in oral fluid using LC/MS/MS. *J Anal Toxicol*. 2007;31(9):596−600.

44. Ngwa G, Fritch D, Blum K, Newland G. Simultaneous analysis of 14 benzodiazepines in oral fluid using by solid phase extraction and LC-MS-MS. *J Anal Toxicol*. 2007;31:369−376.

45. Melanson SE, Griggs D, Bixho I, Khaliq T, Flood JG. 7-aminoclonazepam is superior to clonazepam for detection of clonazepam use in oral fluid by LC-MS/MS. *Clin Chim Acta*. 2016;455:128−133. Available from: https://doi.org/10.1016/j.cca.2016.01.027.

46. Vindenes V, Strand DH, Koksæter P, Gjerde H. Detection of nitrobenzodiazepines and their 7-amino metabolites in oral fluid. *J Anal Toxicol*. 2016;40(4):310−312. Available from: https://doi.org/10.1093/jat/bkw020.

47. Moore C, Rana S, Coulter C. Determination of meperidine, tramadol and oxycodone in human oral fluid using solid phase extraction and gas chromatography-mass spectrometry. *J Chromatogr (Biomed Applns)*. 2007;850:370−375.

48. Hsu YC, Chen BG, Yang SC, et al. Methadone concentrations in blood, plasma, and oral fluid determined by isotope-dilution gas chromatography-mass spectrometry. *Anal Bioanal Chem*. 2013;405(12):3921−3928. Available from: https://doi.org/10.1007/s00216-012-6460-2.

49. Concheiro M, Gray TR, Shakleya DM, Huestis MA. High-throughput simultaneous analysis of buprenorphine, methadone, cocaine, opiates, nicotine, and metabolites in oral fluid by liquid chromatography tandem mass spectrometry. *Anal Bioanal Chem*. 2010;398 (2):915−924.

50. Heltsley R, DePriest A, Black DL, et al. Oral fluid drug testing of chronic pain patients. I. Positive prevalence rates of licit and illicit drugs. *J Anal Toxicol*. 2011;35(8):529−540.

51. Cone EJ, DePriest AZ, Heltsley R, et al. Prescription opioids. IV. Disposition of hydrocodone in oral fluid and blood following controlled single-dose administration. *J Anal Toxicol*. 2015;39(7):510−518. Available from: https://doi.org/10.1093/jat/bkv050.

52. Vindenes V, Yttredal B, Oiestad EL, et al. Oral fluid is a viable alternative for monitoring drug abuse: detection of drugs in oral fluid by liquid chromatography-tandem mass spectrometry and comparison to the results from urine samples from patients treated with methadone or buprenorphine. *J Anal Toxicol*. 2011;35(1):32−39.

53. Moore C, Kelley-Baker T, Lacey J. Interpretation of oxycodone concentrations in oral fluid. *J Opioid Manage*. 2012;8(3):161−166.

54. Gray TR, Dams R, Choo RE, Jones HE, Huestis MA. Methadone disposition in oral fluid during pharmacotherapy for opioid-dependence. *Forens Sci Int*. 2011;206(1−3):98−102.

55. Coulter C, Crompton K, Moore C. Detection of phencyclidine in human oral fluid using solid-phase extraction and liquid chromatography with tandem mass spectrometric detection. *J Chromatogr B Analyt Technol Biomed Life Sci*. 2008;863(1):123−128. Available from: https://doi.org/10.1016/j.jchromb.2008.01.012.

56. Rodrigues WC, Catbagan P, Rana S, Wang G, Moore C. Detection of synthetic cannabinoids in oral fluid using ELISA and LC-MS-MS. *J Anal Toxicol.* 2013;37(8):526−533. Available from: https://doi.org/10.1093/jat/bkt067.

57. Mohamed KM, Al-Hazmi AH, Alasiri AM, Ali Mel-S A. GC-MS method for detection and quantification of cathine, cathinone, methcathinone and ephedrine in oral fluid. *J Chromatogr Sci.* 2016;54(8):1271−1276. Available from: https://doi.org/10.1093/chromsci/bmw082.

58. Vallersnes OM, Persett PS, Øiestad EL, Karinen R, Heyerdahl F, Hovda KE. Underestimated impact of novel psychoactive substances: laboratory confirmation of recreational drug toxicity in Oslo, Norway. *Clin Toxicol (Phila).* 2017;55(7):636−644. Available from: https://doi.org/10.1080/15563650.2017.1312002.

59. Castaneto MS, Wohlfarth A, Desrosiers NA, Hartman RL, Gorelick DA, Huestis MA. Synthetic cannabinoids pharmacokinetics and detection methods in biological matrices. *Drug Metab Rev.* 2015;47(2):124−174. Available from: https://doi.org/10.3109/03602532.2015.1029635.

60. Moore C, Crouch D. Oral fluid for the detection of drugs of abuse using immunoassay and LC-MS/MS. *Bioanalysis.* 2013;5(12):1555−1569. Available from: https://doi.org/10.4155/bio.13.115.

61. A comparison of several LC/MS techniques for use in toxicology. <https://www.agilent.com/cs/library/applications/5990-3450EN.pdf>.

62. Newmeyer MN, Concheiro M, Huestis MA. Rapid quantitative chiral amphetamines liquid chromatography-tandem mass spectrometry: method in plasma and oral fluid with a cost-effective chiral derivatizing reagent. *J Chromatogr A.* 2014;1358:68−74. Available from: https://doi.org/10.1016/j.chroma.2014.06.096.

63. Tuyay J, Garnier M, Mak J, Coulter C, Moore C. Differentiation of amphetamine source based on oral fluid chiral analysis: a preliminary investigation S46. In: *Presented at the Society of Forensic Toxicologists Annual Meeting 2014*, Grand Rapids, MI.

64. Coulter C, Garnier M, Moore C. Analysis of tetrahydrocannabinol and its metabolite, 11-nor-d9-tetrahydrocannabinol-9-carboxylic acid, in oral fluid using liquid chromatography with tandem mass spectrometry. *J Anal Toxicol.* 2012;36(6):413−417.

65. Thieme D, Sachs U, Sachs H, Moore C. Significant enhancement of 11-Hydroxy-THC detection by formation of picolinic acid esters and application of liquid chromatography/multi stage mass spectrometry (LC-MS3): application to hair and oral fluid analysis. *Drug Test Anal.* 2015;7(7):577−585. Available from: https://doi.org/10.1002/dta.1739.

66. Vogeser M, Seger C. Pitfalls associated with the use of liquid chromatography−tandem mass spectrometry in the clinical laboratory. *Clin Chem.* 2010;56(8):1234−1244.

# CHAPTER 8

# Interlaboratory Comparison

## INTRODUCTION

In the 1960s and 1970s, which were the formative years of laboratory medicine, medical practitioners understood that a value such as serum glucose might vary between a hospital laboratory and a reference laboratory (later correctly called a chain laboratory). They also understood that the definition of normal values might vary from laboratory to laboratory. The root causes of the variation were the different analytical methods employed and the populations included as "normal." A clinical chemistry laboratory using a specific glucose oxidase method for glucose might state a normal adult range of 74–100 mg/dL,[1] while another laboratory using a less-specific method such as *ortho*-toluidine might state a range 10% higher than that for the glucose oxidase method. As clinical chemistry developed as a discipline, it became more and more obvious that for exact diagnoses and treatment, all glucose methods and laboratories must produce the same result. Later, the need for uniformity became even more apparent in immunoassay methods such as those used for serum and urine cortisol, chorionic gonadotropin, and a plethora of other analytes.

The above need also became obvious in toxicology and therapeutic drug monitoring, which was often carried out in the same laboratories that produced clinical chemistry tests. A laboratory that reported barbiturate results by differential colorimetry had to produce the same results on a given specimen as another laboratory that used immunoassay or chromatography, and vice versa. Now, except under highly unusual, well-documented circumstances, a reportable analyte should be treated only as a result for that analyte at the specified level, i.e., a morphine of 2500 ng/mL should be exactly that without any equivocation founded on methodology. Historically, several approaches have been used to ensure that all laboratories report the same result for the same analyte and matrix in clinical and forensic testing. Such approaches will be discussed in this chapter, much of which is based

Detection of Drugs and Their Metabolites in Oral Fluid. DOI: https://doi.org/10.1016/B978-0-12-814595-1.00008-8

on the authors' hands-on experience rather than the open literature, which is why there are few references.

## COMMERCIAL CONTROLS

Most clinical chemistry and many toxicology laboratories use commercially available controls for the biological matrix and analytes in question for day-to-day production of results. For example, a toxicology laboratory might use a control that contains benzoylecgonine (BZE) at a certain level such as 225 ng/mL in drug-free urine. Because this control has been assayed by the manufacturer for BZE and is used by several other active laboratories in the production of BZE, a laboratory that obtains results in the range of 203−238 ng/mL can feel confident that they are obtaining valid results on a day-to-day basis. The same laboratory that uses the commercial control can feel confident that they will obtain an acceptable interlaboratory proficiency or performance test result (vide infra) for BZE.

Conversely, a toxicology laboratory that uses the previously referenced hypothetical control and obtains negative values on their initial test assays (cutoff = 150 ng/mL) and quantitative values of around 75−90 ng/mL by chromatography−mass spectrometry urgently needs to re-evaluate their methodologies, including calibration and how they handle donor specimens, controls, and calibrators.

Although commercial oral fluid controls are not as readily available as urine controls, once they become abundantly available and are being used by multiple oral fluid drug testing laboratories, there will be a method of maintaining day-to-day quality and obtaining a penalty-free estimate of how regulated interlaboratory comparisons will turn out.

## BLIND CONTROLS

### Internally Submitted

A blind control is a drug or drug metabolite in the same matrix as that used for the donor specimen being assayed or a negative specimen in the same matrix as that used for donor specimens being assayed. The concentration of drug or metabolite and whether the specimen is positive or negative in the blind control is known only to quality assurance personnel and, perhaps, the laboratory director. Whether required by a regulatory agency[1] or as part of a quality assurance program, a blind sample should be inserted anonymously into the routine workload by placing it among the batch specimens to be initially tested (that is, screened) in the accessioning room without allowing testing personnel to know what it is or where it appears in a batch. Certifying personnel need to know the composition of internal blind specimens and where the blind specimen is placed in an initial test batch. For oral fluid drug and metabolite testing, the blind control sample matrix should be neat human oral fluid or human oral fluid in the form found in the laboratory's routine collection device, depending on what is routinely accepted and tested.

To assess initial testing properly, the drug or metabolite level in the blind specimen should be about 125%–150% of the initial test cutoff. A blind specimen that is several times the laboratory's initial test cutoff is almost assured of producing a positive result but in reality does not effectively challenge the laboratory's analytical system. Conversely, a blind specimen that is at or just above the laboratory's cutoff is as likely to produce a false negative as it is the desired true positive. Time- and freeze–thaw-related decomposition can be a major issue for a laboratory's internal blind specimen system, especially when large batches are frozen and used routinely. Degradation-prone analytes such as cocaine and $\Delta^9$-tetrahydrocannabinol (THC) are more likely to cause false negatives than are stable analytes such as morphine and BZE. If a positive internal blind in a batch tests negative, then that should be the cause for a certifying official to reject the entire batch. If the internal blind tests positive, then it should be carried forward into the confirmatory process after re-pouring in order to challenge both the initial testing and confirmatory processes. However, since confirmatory cutoffs are generally lower than initial testing cutoffs, the laboratory may choose to employ a separate set of blind samples to challenge their confirmatory process. If the laboratory includes specimen validity testing (SVT; see Chapter 6) in its oral fluid drug testing process, both negative and positive SVT should be regularly challenged with internal blinds.

## Externally Submitted

As stated above, an internal blind oral fluid specimen that challenges initial testing, confirmatory testing, and, if applicable, SVT is an integral, not just a peripheral part of oral fluid drug testing. However, the internal blind system described above only challenges a limited portion of accessioning and the analytical portion of the drug testing laboratory. If the entire system is to be tested, an externally submitted blind specimen must be submitted. Externally submitted blind specimens can come from an individual or a corporate body separate from the laboratory such as an organization that specializes in submitting blinds and assessing the laboratory's ability to receive, test, and report. External blind specimens also can originate from an individual within the laboratory who has access to blind accounts (an account that looks real to everyone except the reporting area) and can submit them and reduce the received data on a regular (possibly weekly) basis. External blinds must pass through one of the laboratory's commonly used collection sites in order to appear authentic when received by the laboratory's receiving and accessioning areas and be handled as authentic specimens. External blind drug or drug metabolite positives need to be formulated at a higher analyte level than internal blind specimens to account for losses in handling. To challenge certification and reporting, reviewers (certifiers) and reporters need to be kept out of the external blind loop.

Although external blind systems are needed to assess a laboratory's entire system, there can be problems. The following examples are extracted from the authors' experience and presented only as things to be avoided by practicing laboratories. Urine is the matrix in most examples, as that is where the most experience lies at the time of this writing, but the principles are easily extended to oral fluid.

- An external agency submits a urine specimen containing 500 ng/mL of *d*-methamphetamine and 200 ng/mL of amphetamine. The sample screens positive for amphetamines but fails confirmatory testing. Even though the methamphetamine confirms positive, the amphetamine was 195 ng/mL and the entire specimen had to be reported as negative, resulting in regulatory action. The laboratory is exonerated after a lengthy explanatory process.
- The laboratory director submits an external blind urine control that contains only 1100 ng/mL of methamphetamine when the initial test cutoff is supposed to be 1000 ng/mL. The sample screens negative and is reported as negative even though it should have screened positive. The cause was the director's use of *d,l*-methamphetamine when the antibody in the immunoassay was directed against

*d*-methamphetamine. The antibody in the screening test kit had very little activity toward *l*-methamphetamine. Thus, a negative initial test was obtained legitimately. For future blinds, the director used only *d*-methamphetamine for spiking blind samples.

- The laboratory director submits a negative urine which screens and is confirmed positive for morphine with a small but reportable amount of codeine. The director had consumed a poppy-seed bagel before producing the urine used for the blind. Only certified negative urine should be employed for both negative and positive external blind specimens. The same would apply to oral fluid.

- The laboratory quality assurance agent submits exactly 1.0 mL of the laboratory's synthetic oral fluid spiked to 75 ng/mL of codeine in a Quantisal oral fluid collector. The pad is placed exactly as it would be for an authentic donor specimen, and the collector is capped and sealed. However, the specimen is rejected as substituted. No drug or drug metabolite testing was performed on the specimen by the laboratory's own standard operating procedure. The laboratory uses immunoglobulin G (IgG) for SVT. There is no IgG in the laboratory's synthetic oral fluid matrix. The laboratory had adequate cross-reactivity between codeine and morphine to produce a confirmed positive result; however, the rejection precluded any testing. Unless rejection is a desired outcome of the submission of a blind specimen, authentic drug-free human oral fluid should be used for external blinds wherever possible.

- Wishing to assess the abilities of his laboratory, the director submits a blind oral fluid that is the laboratory's certified negative oral fluid supplemented with human serum from the clinical laboratory adjacent to the forensic area. His laboratory reports positive for THC at 9 ng/mL. The serum came from a marijuana user. The THC positive was a total surprise and had to be explained in the director's weekly quality assurance report. All components of a blind should be certified before formulating a specimen for submission.

- A parent cocaine specimen in neat human oral fluid (25 ng/mL) is submitted to the laboratory by the director. The initial test result is negative. Upon examining the blind specimen, the pH is determined to be 8.0 and only BZE is present. The parent cocaine in the sample decomposed to BZE during the submission process. The antibody in the laboratory's initial test kit is directed toward parent cocaine only, with only about 10% cross-reactivity toward BZE.

It has been the authors' experience that drugs and their metabolites can be dissolved in authentic and synthetic oral fluid successfully, especially when pH is controlled properly. However, a protein such as human IgG or an enzyme such as lactate dehydrogenase may or may not produce values commensurate with what was added, even though complete dissolution may appear to be the case. Using biological fluids in which the protein of interest is already present seems to be superior to adding a weighed-in powdered biological.

## PROFICIENCY TESTING

A sample of known matrix is submitted by an external agency for the laboratory to determine whether it is positive or negative in proficiency testing. Usually, if a sample is positive for a drug or drug metabolite, a quantitative result also will be required by the submitter. In addition to quantitative drug and metabolite testing, SVT also may be challenged with quantifiable analytes (e.g., IgG, albumin). After testing is completed within a time frame dictated by the submitter, results are returned to the submitter for data reduction, scoring, and critique. In most cases, the score received by the submitting laboratory will have a major impact on the laboratory's licensure and certification. The following three proficiency testing programs are sources of oral fluid.

### SAMHSA Oral Fluid Pilot PT Program

The Oral Fluid Pilot PT Program is available to laboratories that currently test oral fluid for amphetamines [including methamphetamine and amphetamine chirality, methylenedioxymethamphetamine (MDMA), and methylenedioxyamphetamine (MDA)], opiates (including codeine, morphine, hydrocodone, hydromorphone, oxycodone, oxymorphone, and 6-acetylmorphine), cocaines (parent cocaine and the major metabolite BZE), THC and its major metabolite [11-nor-$\Delta^9$-tetrahydrocannabinol-9-carboxylic acid (THC-COOH), and phencyclidine (PCP)]. The program is funded by SAMHSA (the US Substance Abuse and Mental Health Administration) and administered by RTI International.

The Pilot Program is free and open to any laboratory that currently performs both initial and confirmatory testing. At the time of this writing, SVT is not required, but it may be in the future. Applicants need only to certify that they can perform the required testing at the levels specified by the program and in the time frame required with each submission. The purposes of the Pilot Program are to provide SAMHSA with information on the current state of oral fluid drug testing and to prepare laboratories to meet the Proposed Mandatory Oral Fluid Guidelines.[2] Although there have been several pauses in the Pilot Program, a set of 15 samples in neat human or synthetic oral fluid is sent overnight four times each year. The laboratories do not receive a score; rather, each participant receives a composite of analytes versus quantitative values. Participants also have the opportunity to take part in a webinar that discusses the results.

At the time of this writing, oral fluid collection devices are not challenged. Only neat oral fluid is distributed to participants.

Individuals wishing to apply for participation in the Pilot Program should contact the National Laboratory Certification Program (NLCP) at RTI International by email (NLCP@rti.org), telephone (+1 919-541-7242), or regular mail (NLCP, RTI International, 3040 East Cornwallis Road, P.O. Box 12194, Research Triangle Park, North Carolina 27709−2194).

## College of American Pathologists

The College of American Pathologists (325 Waukegan Road, Northfield, IL 60093−2750; 800−323−4040 or 847-832-7000; Country code: 001) offers an oral fluid drug testing proficiency testing service (Program Code: OFD). The College ships five challenges per shipment with four shipments per year. According to the 2018 catalog,[3] the OFD provides the following challenges with each shipment, as shown in Table 8.1.

## RTI Proficiency Testing Program for Alcohol and Drugs in Oral Fluid

RTI offers an oral fluid proficiency testing program[4] whose samples will include those listed in Table 8.2.

All samples are provided frozen in neat synthetic oral fluid. Sets of five samples each are shipped three times per year. Participants are allowed 2 weeks post-receipt of samples to complete all testing. Once all results are received by RTI, participants are provided within 2 weeks a detailed report that contains the information necessary for a

## Table 8.1 College of American Pathologists OFD Analytes

| Analyte | Challenges/Shipment |
|---|---|
| Amphetamine group | 5 |
| • Amphetamine | 5 |
| • Methamphetamine | 5 |
| • Methylendioxyamphetamine (MDA) | 5 |
| • Methylenedioxymethamphetamine (MDMA) | 5 |
| • Methylenedioxyethylamphetamine (MDEA) | 5 |
| Benzodiazepine group | 5 |
| • Alprazolam | 5 |
| • Diazepam | 5 |
| • Nordiazepam | 5 |
| • Oxazepam | 5 |
| • Temazepam | 5 |
| Buprenorphine | 5 |
| • Buprenorphine and norbuprenorphine | 5 |
| Cocaine and/or metabolite | 5 |
| • Benzoylecgonine (BZE) | 5 |
| • Cocaine | 5 |
| Cannabinoids | 5 |
| • Delta-9-THC | 5 |
| • Delta-9-THC-COOH | 5 |
| Methadone | 5 |
| Opiate group | 5 |
| • 6-Acetylmorphine (6-AM) | 5 |
| • Codeine | 5 |
| • Hydrocodone | 5 |
| • Hydromorphone | 5 |
| • Morphine | 5 |
| • Oxycodone | 5 |
| • Oxymorphone | 5 |
| Phencyclidine (PCP) | 5 |

## Table 8.2 RTI International PT Program Analytes

| | | |
|---|---|---|
| 6-AM | diazepam | methadone |
| alprazolam | ethanol | methamphetamine |
| amphetamine | flunitrazepam | morphine |
| barbiturates | hydrocodone | nordiazepam |
| benzoylecgonine | hydromorphone | oxazepam |
| buprenorphine | lorazepam | oxycodone |
| clonazepam | MDA | phencyclidine |
| cocaine | MDEA | tetrahydrocannabinol |
| codeine | MDMA | zolpidem |

laboratory to compare its performance quantitatively through group statistics. For more information, contact

Francis M. Esposito, PhD, DABFT
Center for Forensic Sciences
919.316.3837
esposito@rti.org
RTI International
3040 E. Cornwallis Road
PO Box 12194
Research Triangle Park, NC 27709−2194, United States

## PERFORMANCE TESTING

Performance testing challenges not only the laboratory's analytical capabilities but also other critical aspects of specimen handling, including receiving, accessioning, and reporting. At the time of this writing, no program addresses the nonanalytical aspects of oral fluid drug testing. However, since laboratories can employ neat oral fluid or oral fluid diluted 1:3 or 1:4 with various buffer/preservatives, performance testing is critical. Such a need must be addressed for oral fluid to succeed in forensic and clinical toxicology.

## SUMMARY

Oral fluid drug and drug metabolite testing can be validated by testing laboratories both internally and externally. Several external proficiency testing programs exist, but, at the time of this writing, none provides performance testing.

## REFERENCES

1. Adeli K, Ceriotti F, Nieuwesteeg M. In: Rifai N, Horvath AR, Wittwer CT, eds. *Reference Information for the Clinical Laboratory in Tietz Textbook of Clinical Chemistry and Molecular Diagnostics*. 6th ed. Elsevier; 2018.

2. Mandatory Guidelines for Federal Workplace Drug Testing Programs, Substance Abuse and Mental Health Services, Department of Health and Human Services, 82 FR 7920-7970; 2017.

3. 2018 Surveys & Anatomic Pathology Education Programs. Northfield, IL: College of American Pathologists; 2017.

4. Proficiency Testing Program for Alcohol and Drugs in Oral Fluid. Research Triangle Park, North Carolina: RTI International; 2017.

# CHAPTER 9

# Individual Analytes, Specimen Handling, Stability, and Other Issues

## INTRODUCTION

Although academic coursework and formal laboratory training should not be underemphasized, the field of toxicology is very much like the field of medicine: experience is key to success. To quote Arnold P. Lehman, MD, PhD (Chief, Division of Pharmacology, US Food and Drug Administration; Honorary President, Society of Toxicology, 1961−62; co-founder of the Society of Toxicology, 1961), "You, too, can become a toxicologist in two easy lessons... each one taking about 10 years."[1] Thus, this chapter is about the experience of the authors, the authors' laboratories, and informal communications from professional colleagues. References are included where possible.

## ORAL FLUID

Urine is the biological matrix most often employed for clinical, post-mortem, and employment-related toxicology. Urine can be used to identify drugs and metabolites; however, it has extremely limited use in the correlation of xenobiotic levels and clinical state. There are some minor specimen-handling issues. Urine may form refractory precipitates, especially upon refrigeration or during freeze−thaw cycles; it may show microbial growth during ambient or refrigerated storage. However, in general, it is an easily handled biological fluid.

To correlate the level of a xenobiotic with clinical status to include the cause of death, blood and blood products such as serum and plasma are used. In general, serum and any xenobiotics in it are stable at refrigerated temperatures and through several freeze−thaw cycles if remixed completely when thawed. An exception to this stability may be seen when the patient or blood donor is receiving anticoagulant therapy; the serum tends to reclot post centrifugation. Except for

Detection of Drugs and Their Metabolites in Oral Fluid. DOI: https://doi.org/10.1016/B978-0-12-814595-1.00009-X

the occasional formation of microclots, possibly due to inadequate mixture with anticoagulant during the blood draw, whole blood and blood plasma tend to be stable at refrigerated temperatures. As a general observation, the most common clinical and forensic laboratory biological matrices, urine and whole blood and its products, are stable and easily handled.

In direct contrast to urine, blood, and blood products, neat oral fluid without any added buffer, stabilizers, or preservatives is, in the authors' experience, more difficult to handle and stabilize. Given several different neat oral fluid samples in separate non-silanized glass tubes, one specimen might produce a solid precipitate, while another might develop a string-like series of (probably) proteinaceous fibers. Other samples may remain crystal clear and colorless and almost as fluid as water for days or even weeks of sitting at ambient temperature on an office table, while a different specimen may form a gel that remains for weeks to months at room temperature. Gels, precipitates, and protein fibers may be broken up and, in some cases, actually be redissolved after minutes to hours of inversion or movement on a lateral motion laboratory shaker. In other cases, a neat oral fluid gel is refractory to redissolution or even breaking up. As shown in Chapter 6 on specimen validity testing, all neat oral fluid samples produce a precipitate of cells and other oral cavity debris regardless of whether the upper fluid layer precipitates or coagulates. Although other permutations have been observed in handling thousands of neat oral fluid samples and mixtures, as a general statement, the form of a neat oral fluid on standing for hours to days is unpredictable. All individual observations aside, the main point is that neat oral fluid as a matrix from which to extract small molecules such as drugs and drug metabolites may be required to demonstrate the true level, but it is unpredictable in many instances.

The solution for the above unpredictability is to dilute oral fluid with a buffer or other preservative solution, which has caused the development of many oral fluid collection devices. Such dilution, which generally leads to the elimination of gelling and the formation of precipitates and enhanced stabilization of the drugs and their metabolites, can be accomplished by either direct dilution or collection of oral fluid onto a fibrous pad which is then placed into a buffer/preservative (see Chapter 3 on sample collection).

Where a large amount of neat oral fluid is required, collection may be performed by expectoration into a plastic receptacle such as a Falcon tube while maintaining the tube on wet ice. Post-collection, the sample may be frozen until use. For routine drug testing, the use of a collection device with buffer is preferred for accurate, valid drug quantification.

## Δ9-TETRAHYDROCANNABINOL

In forensic and clinical urine toxicology, the main cannabinoid measured is 11-nor-$\Delta^9$-tetrahydrocannabinol-9-carboxylic acid (THC-COOH, also called THCA). As has been stated in many publications,[2] THC-COOH is unstable and may be lost from both biological and synthetic matrices by a plethora of mechanisms. However, even though the stability and preservation of THC-COOH have been a challenging issue for test kit manufacturers and many forensic and clinical laboratories, the parent THC ($\Delta^9$-tetrahydrocannabinol) is even more unstable and subject to oxidative degradation, chemical rearrangement, and disappearance from various matrices due to purely physical phenomena such as surface adsorption and absorption.

We have found that unmodified parent THC can be dissolved in neat synthetic oral fluid which contains a small amount of protein. When THC-containing neat oral fluid was placed in silanized amber glass vials, frozen, and shipped at ambient temperature, parent THC was stable enough to be useful for initial and quantitative confirmatory testing in laboratories to which the samples were sent as part of the US Substance Abuse and Mental Health Services Administration (SAMHSA) Oral Fluid Pilot Proficiency Testing (PT) Program (see Chapter 8). In the more recent SAMHSA Pilot Program offerings, parent THC under the conditions previously described and in modified neat human oral fluid was found stable enough to be useful over the 2-week period that participants were given in which to analyze and report sample results.[3] Conversely, when parent THC was spiked into the unmodified neat oral fluid of two donors in a drug study, substantial loss of THC was observed in samples sent to participant laboratories.[4] As of this writing, the mechanism of the loss of parent THC from the two unmodified neat human oral fluid samples is unknown, with pH not being suspect in either case. By itself, parent THC appears to be most stable around pH 7.5.

THC in expectorated oral fluid is less stable than THC in device buffers.[5] It is also not stable even in Quantisal buffer/preservative under fluorescent lights, with losses greater than 50% reported over 14 days. In the dark at room temperature the losses were only 20%.[6]

While laboratory analysis of THC in oral fluid is routine, with both immunoassays and confirmatory mass spectrometry methods widely available, the storage conditions for each device recommended by the manufacturer for drug positive samples may be different.

It has been suggested that THC-COOH may indicate cannabis use as opposed to passive exposure. If its analysis becomes routine in forensic or clinical laboratories, the need for deconjugation of the glucuronide conjugate, which is about two-thirds of the total THC-COOH, may need to be considered.[7]

Although THC usually is spiked into oral fluid calibrators and controls separately from other drugs, please see the Amphetamines section below for a comment on changes in THC when its solutions are overacidified.

## COCAINE

Parent cocaine (pharmacologically active) and its major metabolite, benzoylecgonine (BZE) (pharmacologically inactive), are both of interest in oral fluid drug testing. Both have a basic nitrogen in the carbon bridge. Other metabolites such as ecgonine methyl ester, norcocaine, ecgonine itself, and minor oxidation products of cocaine and BZE may be of interest in some special studies.

Parent cocaine is a double ester while BZE has only a single ester moiety. In addition to an ester group, BZE has a carboxylic acid function and still retains the basic bridge nitrogen, making it zwitterionic. Cocaine has very little solubility in water (1 g dissolves in 600 mL at room temperature[8]) while BZE has considerable water solubility (the tetrahydrate can be crystallized from water, in which it is soluble when the water is hot). Thus, the parent cocaine is extracted easily from aqueous matrices, especially when the analyst makes the matrix mildly basic. Conversely, BZE is more difficult to extract from biological matrices by liquid–liquid technology and requires moderately polar solvents such as methylene chloride and chloroform. Cocaine readily decomposes to BZE even at neutral pH, and more rapidly in strong base and strong acid, as is typical of most organic esters.[9] In aqueous solution, parent cocaine appears to demonstrate its greatest stability around pH 5. In the presence of ethyl alcohol (ethanol), cocaine may transesterify to the pharmacologically active substance cocaethylene.

Based on the above discussion, the stabilization of parent cocaine in a matrix such as oral fluid, which contains esterases possibly derived from crevicular fluid, is challenging unless appropriate buffering at a pH where parent cocaine is stable and is employed. Surprisingly, BZE appears stable in most matrices at a broad range of pH values.

Providing the combination of parent THC and parent cocaine in the same synthetic or human oral fluid can usually be accomplished with a compromise pH of approximately 6.5. It is notable that the pH of a spiking or stock solution also must be considered. The addition of drugs such as amphetamines or opiates (vide infra) to an unbuffered spiking solution can cause an unanticipated rise in pH, resulting in the hydrolysis of parent cocaine and 6-acetylmorphine.

Ventura et al. reported on the stability of drugs in transit for the purpose of assessing external proficiency schemes in laboratories. They evaluated two different collection devices and noted that 26%−41% of the cocaine degrades to BZE within 48−72 h, respectively.[10]

# AMPHETAMINES

In general, the stereoisomers of amphetamine, methamphetamine, methylenedioxymethamphetamine (MDMA, "ecstasy"), methylene-dioxyamphetamine (MDA), and methylenedioxyethylamphetamine (MDEA) are common in oral fluid toxicology. The literature is replete with many other amphetamine derivatives that possess stimulatory and/or hallucinogenic capability and may be detected and, if necessary, quantitated in human oral fluid.

The fundamental amphetamines, amphetamine itself, and metham-phetamine are basic (amphetamine[11] and methamphetamine[12] $pK_a = 9.9$). Both exist in two stereoisomeric forms. The $d$ and $l$ forms of amphetamine and the $d$ form of methamphetamine have stimulatory activity. The $l$ form of methamphetamine is active only as an antihista-mine.[13] Chiral analysis of methamphetamine in oral fluid can assist in distinguishing licit use (Vicks inhaler, *l* form) from illicit use (*d* form); for amphetamine, a 3:1 ratio of *d* to *l* amphetamine may indicate Adderall ingestion (see Chapter 7).

As stated above, it is important to consider the pH of a spiking solution or finished oral fluid control or calibrator when amphetamines are added and parent cocaine and/or 6-acetylmorphine is present. A seemingly innocuous elevation of the pH of a spiking solution or a finished synthetic or human oral fluid product can be disastrous for identifying cocaine and 6-acetylmorphine. Overcompensation for a pH elevation by reducing pH to less than 4 can result in an unexpected significant reduction of $\Delta^9$-THC due to isomerization to $\Delta^8$-THC if $\Delta^9$-THC is in the same spiking mixture.[14]

Although very few stability issues have arisen with the methylenedioxy drugs, it is worth noting that the methylenedioxy ring is the ether form of a geminal diol. Geminal diols[15] are noted for their instability, which in the case of the methylenedioxy compounds is greatly stabilized by being present as a diether but may be a potential area for unwanted chemical reactions.

In the above discussion, it is again noteworthy that hydrochloride salts may contribute unwanted acidity and lowering of the pH of a spiking solution. Conversely, use of the free base may result in an undesirable rising of the pH of a spiking solution.

## OPIATES

Although many chemical modifications can be made to the fundamental opiates codeine and morphine, in general they are resistant to chemical changes caused by mild oxidation and reduction during routine extraction and, where appropriate, derivatization. Despite morphine's polarity and mildly amphoteric nature, both codeine and morphine can be extracted routinely from oral fluid and analyzed.

In urine, to obtain total codeine and total morphine concentrations, deconjugation is required. The same requirement applies to blood and its products unless only free active drug concentration is desired. Whether deconjugation will be required for the analysis of codeine and morphine in oral fluid is unknown at the time of this writing.

Like the analysis of codeine and morphine, analysis of the semisynthetic opiates hydrocodone, hydromorphone, oxycodone, and oxymorphone is relatively unrestricted. In contrast to urinalysis for oxycodone use, the parent drug itself is predominantly identified in oral fluid, and the main metabolite is noroxycodone and not oxymorphone. Likewise, norcodeine is present after codeine intake (not morphine) and norhydrocodone (not hydromorphone) after hydrocodone intake. A comprehensive review of the metabolism of prescription opioids, including codeine, hydrocodone, oxycodone, fentanyl, meperidine, methadone, buprenorphine, and tramadol, was published in 2015.[16] The authors noted that oral fluid-to-blood ratios exceed 1 for most of these opioids, making oral fluid an excellent alternative matrix for testing of this drug class.

Two opiates which require special preanalytical and analytical attention are heroin (diacetylmorphine) and 6-monoacetylmorphine (6-AM), which is the intermediate metabolite of diacetylmorphine. Both are labile esters. However, both molecules are stabilized at about pH 5. It appears from experience that common collection pH of 6.5 is adequate to stabilize 6-AM.[3] In a transit study, Ventura et al. noted that 9%−12% of 6-AM converted to morphine within 48 h.[10]

Pain management and prescription medication compliance are areas where the utility of oral fluid testing is gaining wide acceptance. Several publications indicate similar or better detection rates for drugs in oral fluid compared to urine.[17,18] The interpretation of oral fluid opioid concentrations for therapeutic and forensic purposes is also gaining traction and will be an area of future research.[19,20]

## BENZODIAZEPINES

As a class of compounds, the 1,4-benzodiazepines are pharmacodynamically potent but chemically unstable. They demonstrate pH, light, and oxidation sensitivity. Perhaps, the only two stable ones for immunoassay calibration and control are nitrazepam (not approved for use in the United States) and nordiazepam (metabolite of diazepam).

The amino-metabolites of flunitrazepam, nitrazepam, and clonazepam are present at higher concentration and are more stable than the parent drugs in oral fluid.[21] From 1001 samples positive for clonazepam and its main metabolite 7-aminoclonazepam, both were detected in 70.6%, only clonazepam in 6.3%, and only the metabolite in 23% of the specimens. For nitrazepam the numbers were similar: both drug and metabolite were present in 65.8% of specimens, the parent drug only in 7.5%, and metabolite only in 26.5%.[22]

# REFERENCES

1. Lehman AP. Perspectives on Risk Assessment. In: SOT FDA Colloquia on Emerging Toxicological Science: Challenges in Food and Ingredient Safety; 2015.

2. White RM. Forensic Instability and poor recovery of cannabinoids in urine, oral fluid, and hair. *Forensic Sci Rev*. 2018;30:33−49.

3. White RM, Hart ED, Mitchell JM. SAMHSA Oral Fluid Pilot PT; 2011−2017.

4. White RM, Sutheimer C, Mitchell JM, Hart ED, Weber FX, Lodico C. The influence of zinc-containing over-the-counter products and an oral rinse on oral fluid drug testing. Submitted for publication.

5. Lee D, Milman G, Schwope DM, Barnes AJ, Gorelick DA, Huestis MA. Cannabinoid stability in authentic oral fluid after controlled cannabis smoking. *Clin Chem*. 2012;58 (7):1101−1109.

6. Moore C, Vincent M, Rana S, Coulter C, Agrawal A, Soares J. Stability of Δ9-tetrahydrocannabinol (THC) in oral fluid using the Quantisal collection device. *Forens Sci Int*. 2006;164:126−130.

7. Moore C, Rana S, Coulter C, Day D, Vincent M, Soares J. Detection of conjugated 11-nor-Δ9-tetrahydrocannabinol-9-carboxylic acid in oral fluid. *J Anal Toxicol*. 2007;31(4):187−194.

8. *The Merck Index*, 15th ed. Cambridge, United Kingdom: The Royal Society of Chemistry; 2013.

9. Brown WH, Foote CS, Iverson BL, Anslyn EV, Novak BM. *Functional Derivatives of Carboxylic Acids. Organic Chemistry*. Belmont, CA: Brooks/Cole; 2012.

10. Ventura M, Pichini S, Ventura R, et al. Stability of drugs of abuse in oral fluid collection devices with purpose of external quality assessment schemes. *Ther Drug Monit*. 2009;31 (2):277−280. Available from: https://doi.org/10.1097/FTD.0b013e318198670b.

11. Amphetamine. In: Baselt RC, ed. *Disposition of Toxic Drugs and Chemicals in Man*, 10th ed., Seal Beach, CA: Biomedical Publications; 2014, p. 122.

12. d-Methamphetamine. In: Baselt RC, ed. *Disposition of Toxic Drugs and Chemicals in Man*, 10th ed., Seal Beach, CA: Biomedical Publications; 2014, p. 1263.

13. l-Methamphetamine. In: Baselt RC, ed. *Disposition of Toxic Drugs and Chemicals in Man*, 10th ed., Seal Beach, CA: Biomedical Publications; 2014, p. 1266.

14. Garrett ER, Tsau J. Stability of the tetrahydrocannabinols I. *J Pharm Sci*. 1974;63:1563.

15. Brown WH, Foote CS, Iverson BL, Anslyn EV, Novak BM. *Aldehydes and Ketones. Organic Chemistry*. Belmont, CA: Brooks/Cole; 2012.

16. DePriest AZ, Puet BL, Holt AC, Roberts A, Cone EJ. Metabolism and disposition of prescription opioids: a review. *Forensic Sci Rev*. 2015;27(2):115−145.

17. Kunkel F, Fey E, Borg D, Stripp R, Getto C. Assessment of the use of oral fluid as a matrix for drug monitoring in patients undergoing treatment for opioid addiction. *J Opioid Manage*. 2015;11(5):435−442. Available from: https://doi.org/10.5055/jom.2015.0293.

18. Conermann T, Gosalia AR, Kabazie AJ, et al. Utility of oral fluid in compliance monitoring of opioid medications. *Pain Phys*. 2014;17(1):63−70.

19. Moore C, Kelley-Baker T, Lacey J. Interpretation of oxycodone concentrations in oral fluid. *J Opioid Manage.* 2012;8(3):161–166. Available from: https://doi.org/10.5055/jom.2012.0112.

20. Shaparin N, Mehta N, Kunkel F, Stripp R, Borg D, Kolb E. A novel chronic opioid monitoring tool to assess prescription drug steady state levels in oral fluid. *Pain Med.* 2017;18 (11):2162–2169. Available from: https://doi.org/10.1093/pm/pnw335.

21. Melanson SE, Griggs D, Bixho I, Khaliq T, Flood JG. 7-aminoclonazepam is superior to clonazepam for detection of clonazepam use in oral fluid by LC-MS/MS. *Clin Chim Acta.* 2016;455:128–133. Available from: https://doi.org/10.1016/j.cca.2016.01.027.

22. Vindenes V, Strand DH, Koksaeter P, Gjerde H. Detection of nitrobenzodiazepines and their 7-amino metabolites in oral fluid. *J Anal Toxicol.* 2016;40(4):310–312. Available from: https://doi.org/10.1093/jat/bkw020.

# CHAPTER 10

# Concluding Remarks

Analysis of oral fluid for drugs and their metabolites is of increasing interest in many settings, including pain management, medication adherence, workplace drug testing, driving under the influence cases, and probation and parole programs. This book, while by no means exhaustive, is focused on the advantages and disadvantages of using oral fluid as a drug testing matrix. Improvements in collection devices, homogeneous immunoassays, and the sensitivity of bench-top confirmation instrumentation combined with a deeper understanding of pharmacokinetics and drug disposition in oral fluid have contributed to the expansion of use. In a trend towards parity with urine as the matrix of choice, professional groups have begun to recommend cutoff concentrations for specific areas of testing, and to prepare and promote performance testing programs to assist laboratories in achieving reliable and consistent quantitative results.

## STRENGTHS OF ORAL FLUID AS A TEST MATRIX

- Preferred specimen type by donors.
- Easy, minimally invasive, rapid collection.
- Detects recent intake of drugs and alcohol when tested at standard cutoffs.
- Ideal for post-accident and for-cause drug testing.
- Potential for therapeutic interpretation of drug concentrations.
- Proficiency programs available.

## WEAKNESSES OF ORAL FLUID AS A TEST MATRIX

- Shorter window of drug detection than urine or hair.
- Not ideal for preemployment drug testing.
- Collection of a second "split" sample not as straightforward as for urine.

Detection of Drugs and Their Metabolites in Oral Fluid. DOI: https://doi.org/10.1016/B978-0-12-814595-1.00010-6

## METHODOLOGICAL NEEDS

- FDA clearance for assays using workplace drug testing cutoffs.
- Devices which collect adequate oral fluid to split into two separate specimens.
- Immunoassay and routine confirmation procedure for the detection of marijuana metabolite THC-COOH at relevant concentrations.

## CUTTING-EDGE TRENDS

- Better interpretation of oral fluid drug concentrations for clinical use.
- Development of algorithms which include pharmacokinetic and pharmacodynamic parameters to assist in oral fluid drug concentration interpretation.
- Development of predictive oral fluid drug concentrations which identify an individual as being "under the influence" of a specific drug.

## SUMMARY

Fundamentally, a quantitative concentration of an active drug in saliva, which becomes the actually sampled oral fluid, will have some degree of correlation with a corresponding whole blood drug level. However, the correlation is strongly dependent on saliva pH and other factors involved in the blood-to-saliva transfer, so an oral fluid concentration may not reflect a whole blood concentration directly. Even so, the presence or absence of an active drug in oral fluid is a far better indicator of recent use than in urine. Thus, a major strength of a properly performed oral fluid drug test lies in the criminal law fields of driving under the influence and drug-facilitated crimes, notably in the administrative areas of post-accident and for-cause testing. Civil cases, which many times are an outgrowth of criminal and administrative cases, also may be influenced or even decided by oral fluid drug test results.

Whether neat oral fluid or fluid collected by a pad device is used, the analytical toxicologist must recognize that collection, transportation, accessioning, testing, and storage—that is, methods and sample flow—differ widely from those for urine and blood or blood products.

This does not mean that oral fluid has a fatal flaw, but rather that the differences must be recognized in validated analytical procedures.

Regardless of whether neat oral fluid or a pad-type collector is used, all sample collections should be observed. Even though specimen validity testing may not have as important a role in oral fluid testing as it does in urine testing, it still needs to be given strong consideration to ensure proper sample collection technique.

Currently, the analytical methodologies for laboratory-based initial and confirmatory testing exist. Point of collection testing methods are also available but probably need augmented sensitivity for some drug classes, specifically benzodiazepines.

Because oral fluid analytes and the matrix itself differ from those of urine and blood, performance testing and other interlaboratory comparisons will have different requirements that need to be worked out before a system is put into operation. It is a relatively simple task for blind positive, negative, or altered urine to be introduced into a laboratory quality system but more complex to do so for oral fluid. The issues apply to any organization that submits blind samples, such as the US Department of Transportation.

At the time of this writing, oral fluid is used primarily for employment-related, probation, and parole testing. It is ready for wider application.

# INDEX

*Note*: Page numbers followed by "*f*" and "*t*" refer to figures and tables, respectively.